NAROSA SERIES IN POWER AND ENERGY SYSTEMS

Electrical Engineering Materials

NAROSA SERIES IN POWER AND ENERGY SYSTEMS

Series Editor

S.C. SRIVASTAVA
Department of Electrical Engineering
Indian Institute of Technology Kanpur
Kanpur, India

Series Titles:

ELECTRIC POWER DISTRIBUTION AUTOMATION
R.P. Gupta

ELECTRICAL ENGINEERING MATERIALS
Nava Raj Karki, Bhadra Prasad Pokharel

OPERATION AND MANAGEMENT OF POWER SYSTEM IN ELECTRICITY MARKET
S.C. Srivastava, S.N. Singh

POWER ELECTRONICS LABORATORY Theory, Practice & Organization
O.P. Arora

REACTIVE POWER ANALYSIS OF ISOLATED HYBRID POWER SYSTEMS
R.C. Bansal

RESTRUCTURED POWER SYSTEMS
S.A. Khaparde, A.R. Abhyankar

SMALL SIGNAL ANALYSIS OF POWER SYSTEMS
M.A. Pai, D.P. Sen Gupta, K.R. Padiyar

NAROSA SERIES IN POWER AND ENERGY SYSTEMS

Electrical Engineering Materials

Bhadra Prasad Pokharel
Nava Raj Karki

Alpha Science International Ltd.
Oxford, U.K.

Bhadra Prasad Pokharel
Nava Raj Karki
Institute of Engineering
Tribhuvan University
Lalitpur, Nepal

Alpha Science International Ltd.
7200 The Quorum, Oxford Business Park North
Garsington Road, Oxford OX4 2JZ, U.K.

www.alphasci.com

ISBN-13: 978-1-84265-297-8
ISBN-10: 1-84265-297-4

Printed in India

Dedicated
to
Our Parents

Foreword

Recent technological developments in all fields of engineering are largely attributed to the advancements in the required materials. Continued efforts in the invention of new materials having better insulating, conducting, semiconducting and magnetic properties have brought revolutionary changes in the design and development of more efficient heavy electrical equipments as well as miniaturized electronic components and circuits. In view of this, it is important for an engineering graduate to learn the basic properties of various materials used and their applications.

This book has addressed this requirement of electrical, electronics and computer engineering students by covering the basic properties, theoretical principles and applications of all types of electrical engineering materials. Although the book has been written for the undergraduate engineering students, it will be quite useful to the graduate engineering science students, practicing Engineers as well as Physicists.

The authors, Mr. Nava Raj Karki and Dr. Bhadra Prasad Pokharel, faculty members of Pulchowk Campus, Institute of Engineering, Tribhuvan University, Nepal should be commended for this valuable and well written book, being published by the Narosa Publishing House, New Delhi. I am sure that the engineering students and graduates, will find this book quite useful.

S. C. Srivastava
Department of Electrical Engineering
Indian Institute of Technology Kanpur, India

Preface

The students of undergraduate level courses in electrical, electronics and computer engineering require fundamental knowledge on materials that are used to manufacture electronics, electrical and computer devices, machines, instruments and equipments. To understand the operation of these devices and constructional requirements, it is essential to have sound knowledge of the materials that are used to manufacture the above mentioned devices and equipments. The selection of materials for the manufacture of any electrical/electronics device is based on what the expected output is from the device. To achieve this, in-depth knowledge of the properties of the materials used is a must. The main aim of the course "Electrical Engineering Materials" at undergraduate level is to teach the students of electrical engineering discipline the basic electrical and magnetic properties of the materials.

The lack of a single textbook, which covers the whole content, taught at undergraduate level in electrical, electronics and computer engineering programs has prompted the authors to write this book. The experience gained by the authors while leading the same course at Institute of Engineering and other Campuses has greatly assisted to spot the weaker side of students and hence the authors have tried to prepare this manuscript to help them out.

The authors are thankful to Prof Dr S C Srivastava, Department of Electrical Engineering, Indian Institute of Technology Kanpur, India for kindly consenting to write a Foreword. The authors are also thankful to the students at undergraduate level at various engineering colleges who patiently listened to the lectures delivered to them without which the authors might not have gained ample experience to start this venture.

Bhadra Prasad Pokharel
Nava Raj Karki

Contents

Theory of Metals

Chapter 1

1.1 Elementary Quantum Mechanical Ideas

Quantum mechanics is a generalized form of mechanics applicable to very small objects like electrons in the atom. Large systems can also be described by quantum mechanics but mostly a much simpler form of mechanics-classical mechanics is used to describe them. Many things that we have experienced throughout our life are not applicable to quantum mechanics. Rather some strange phenomena (in the sense of common man's understanding) are inherent to quantum mechanics. So when studying quantum mechanics it might seem during the course we may get lost out of realm but still will be forced, based upon immense works and results of these works by many scientist, to agree with these findings.

Various processes and phenomena that take place within a material cannot be explained without having good understanding of quantum mechanics. Although classical mechanics gives us information about various behaviors of different materials, but it fails to explain the material behaviors when it comes to microscopic level. This is the region where quantum mechanics fits the bill, i.e. quantum mechanics is able to explain this or that phenomenon.

1.1.1 Wave Particle Duality: De Broglie's Equation

During the early part of twentieth century scientists failed to explain number of phenomena by using Newton's laws of motion and Maxwell's electromagnetic theory. It was discovered that particles of atomic dimensions sometimes behave more like waves than discrete classical particles. It was also observed that electromagnetic waves, e.g., x-rays, gamma rays, visible light etc sometimes exhibit properties similar to properties of discrete particles of matter.

Experiments showed that particles like electrons, protons and neutrons exhibit an equivalent wavelength given by

$$\lambda = \frac{h}{p} \text{ De Broglie's Equation} \tag{1.1}$$

Where λ = equivalent wavelength exhibited by the particle

p = momentum of the particle

h = Planck's universal constant

$= 6.624 \times 10^{-34}$ J.s

Also, we can write the above equation in the following form

$$p = \hbar k \tag{1.2}$$

where $\hbar = h/2\pi$ and $k = 2\pi/\lambda$ wave number

Other experiments showed that electromagnetic radiation energy is sometimes contained in "***packets***" known as **photons**.

$$E = hf \tag{1.3}$$

Equation (1.3) is called Planck's equation.

Frequency of electromagnetic radiation f is related to radian frequency ω by

$$\omega = 2\pi f \quad \text{or} \quad f = \omega/2\pi$$

Putting the value of f in equation (1.3), we will get

$$E = \hbar\omega \tag{1.4}$$

There were two famous experiments conducted early in the twentieth century that were unexplained by classical mechanics and so warranted the need for quantum mechanical ideas.

Experiment 1:

It was observed that beams of particle such as electrons were diffracted from a crystal in a manner similar to the diffraction of x-rays, which are electromagnetic waves.

$$E = \frac{mv^2}{2} = \frac{p^2}{2m} \tag{1.5}$$

$$P = (2\text{mE})^{1/2}$$

Experiment 2:

When electromagnetic radiation interacts with electrons in a solid or gas, a certain maximum wavelength of radiation is required to eject electrons from the surface of the material. By bombarding the material with photons of different energies it is found that for each material a characteristic energy is required to remove the electrons from the surface of the material. The minimum frequency (hence the maximum wavelength) of this electromagnetic radiation was found to be

$$f = \frac{\phi}{h} \tag{1.6}$$

where Φ = characteristic energy known as work function of the material.

Comparing equations (1.3) and (1.6), we can conclude that the electromagnetic radiation behaves like particles with packets of energy when it collides with microscopic particles like electrons in the material.

From above discussed we can clearly see the duality of wave and particle. Particles like electrons sometimes behave like wave. And wave like electromagnetic radiation sometimes behaves like particles (photons with packets of energy). In quantum mechanics there are no clear distinctions between wave and particle – waves sometimes exhibiting particle like properties and particles sometimes exhibiting wave-like properties. So, what is this- wave or particle? Till now there is not a single name that is commonly used to represent this wave-particle duality. Some say it "**wavicle**", some say it "**particle-wave**" and some say it "**wave-particle**". Whatever be the nomenclature used to represent this duality, the duality itself exists and is of utmost importance compared to its nomenclature.

1.1.2 Wave Function: Schrödinger's Equation

Since there is no clear distinction between wave and particle in quantum mechanics, there should be some clear representation (maybe in the form of mathematical expression) of what we call wave-particle duality.

In quantum mechanics there exists an expression in the form of equation that is used to represent the wave-particle duality.

Similar to ordinary wave, the displacement about its mean position is represented by wave-function ψ as

$$\psi = Ae^{-j\left(\omega t - \frac{2\pi}{\lambda}x\right)}$$

Using $$E = hf = 2\pi\hbar f = \frac{2\pi\hbar c}{\lambda}$$

i.e. $$\lambda = \frac{h}{p} = \frac{2\pi\hbar}{p}$$

$$\psi = Ae^{-\frac{j}{\hbar}(Et - px)} \tag{1.7}$$

where A = maximum displacement from mean position

λ = wave length

x = distance from origin

f = frequency of vibration

ω = radian frequency

$k = 2\pi/\lambda$ = wave number

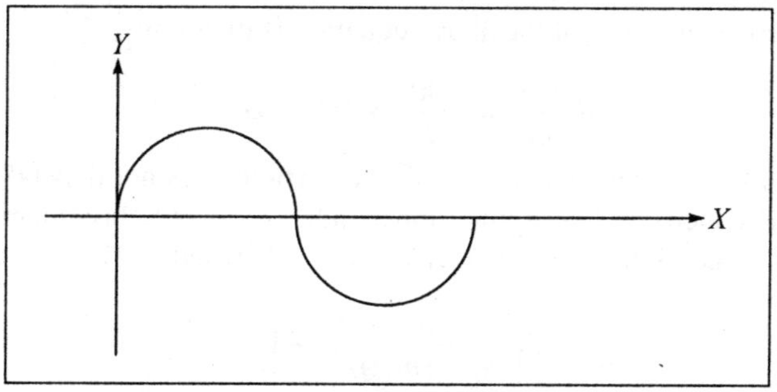

Fig. 1.1 Ordinary wave represented by equation (1.7)

Equation (1.7) is a mathematical description of the wave equivalent of an unrestricted particle with total energy E and momentum p moving in the +x direction, and is correct only for freely moving particles.

Schrödinger's equation is the fundamental equation of quantum mechanics and is also wave equation in the variable ψ.

Differentiating equation (1.7) twice with respect to x, we will get

$$\frac{d\psi}{dx} = Aj\frac{2\pi}{\lambda}e^{-j\left(\omega t - \frac{2\pi x}{\lambda}\right)}$$

Then again differentiating w.r.t. x, we get

$$\frac{d^2\psi}{dx^2} = Aj^2k^2e^{-j\left(\omega t - \frac{2\pi x}{\lambda}\right)} = -\frac{p^2}{\hbar^2}\psi \tag{1.8}$$

And, differentiating equation (1.7) w.r.t. time, we get

$$\frac{d\psi}{dt} = -j\frac{E}{\hbar}Ae^{-\frac{j}{\hbar}(Et-px)}$$

$$\frac{d\psi}{dt} = -j\frac{E}{\hbar}\psi \qquad (1.9)$$

At speeds small compared to that of light, the total energy E of a particle is the sum of its KE ($p^2/2m$) and potential energy PE(V) where V is in general a function of position x and time t.

$$E = KE + PE$$

or,

$$E = \frac{p^2}{2m} + V \qquad (1.10)$$

Multiplying both sides by wave-function ψ, we get

$$E\psi = \frac{p^2}{2m}\psi + V\psi \qquad (1.11)$$

Then from equations (1.8) and (1.9), we get

$$j\hbar\frac{\partial\psi}{\partial t} = -\frac{\hbar^2}{2m}\frac{\partial^2\psi}{\partial x^2} + V\psi \qquad (1.12)$$

Equation (1.12) is the time-dependent form of Schrödinger's wave equation in one –dimension. In 3-dimension, the time dependent form of the above equation is given by

$$j\hbar\frac{\partial\psi}{\partial t} = -\frac{\hbar^2}{2m}\nabla^2\psi + V\psi \qquad (1.13)$$

It has been observed that in many situations, V of a particle does not depend on time and varies only with its position only. For such conditions, time-independent form of Schrödinger's equation is applicable which can be obtained as mentioned below. From equation (1.9) and (1.12), we get

$$E\psi = -\frac{\hbar^2}{2m}\frac{\partial^2\psi}{\partial x^2} + V\psi \qquad (1.14)$$

$$\frac{\partial^2\psi}{\partial x^2} + \frac{2m}{\hbar^2}(E - V) = 0 \qquad (1.15)$$

Equation (1.15) is the steady-state form of Schrödinger's equation in one-dimension. In 3-dimension, it can be expressed as

$$\nabla^2\psi + \frac{2m}{\hbar^2}(E - V)\psi = 0 \qquad (1.16)$$

where $\nabla^2 = \frac{\partial^2}{\partial x^2} + \frac{\partial^2}{\partial y^2} + \frac{\partial^2}{\partial z^2}$ is called Laplacian operator.

Solving the above differential equations (1.12), (1.13), (1.15) and (1.16) we get a wave function ψ which fulfills the requirements for an acceptable wave-function, i.e., its derivatives be continuous, finite and single valued.

1.1.3 Wave Function ψ

The wave in classical mechanics can easily be understood and represented. There are numerous analogy to the wave in classical mechanics. Unlike in classical mechanics, the wave function does not enjoy the privilege of having similar analogies in quantum mechanics that will enable us to easily understand the wave function. Despite the difficulties in finding analogies to describe the wave function in quantum mechanics, it is used to describe what is called an "observable" in quantum mechanics.

In quantum mechanics, energy, momentum and position of the particle are called observables and the wave function ψ is used to describe these observables. These observables which characterize the particle can not be measured at the same instant, i.e., only one can be measured at once.

The equation (1.15) and (1.16) were derived assuming that all the physical entities exhibit some wave properties, i.e., all the particles behave like waves. The wave function is extensively used to predict the probability that a particle would be found in a specified volume of space.

The wave function ψ is, in general, a complex three dimensional quantity (i.e., is the function of all three coordinates) and is dependent upon time. To be precise the wave function has both the real and imaginary components that will characterize the particle in quantum mechanical domain.

If $\psi(x, y, z, t)$ represents wave function, then $|\psi|^2$ is called probability amplitude.

Let dxdydz be an elementary volume and the probability of the particle represented by wave function ψ being located in this incremental volume is to be assessed. The probability of finding the particle in given incremental volume is equal to the product of probability amplitude (also called probability density function) and the incremental volume, i.e.

$$|\psi(x, y, z)|^2 dxdydz$$

If we have to find out the probability of locating the particle between x and x+dx in single dimensional case, then the probability will be equal to

$$|\psi(x)|^2 dx$$

But if the particle is located within the given interval (from x to x+dx in single dimensional case), the sum of all probabilities that the particle is to be found in the above mentioned interval must be equal to 1. Mathematically, for a particle located somewhere in the universe, we can write

$$\int_{-\infty}^{+\infty} \psi(x, t)dx = 1 \tag{1.17}$$

1.1.3.1 Hypothetical example

When we conducted various measurements in a system (fictitious not real), the probability amplitude curve vs. position of electron was found to be as shown in fig. 1.2 below. From the figure below it is clear that the electron is always found between x_1 and x_4. So, the sum of all probabilities that the electron would be found between x_1 and x_4 is equal to unity. And, the probability of finding the electron beyond these limits is equal to zero.

Mathematically, we can write

$$\int_{-\infty}^{+\infty} |\psi(x)|^2 \, dx = 1 \tag{1.18}$$

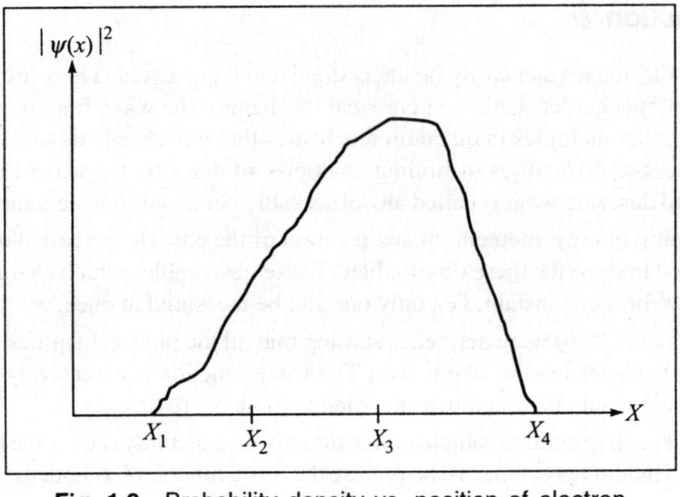

Fig. 1.2 Probability density vs. position of electron

From the Fig. 1.2 it can be said that the particle is most likely to be found in the vicinity of x_3. The probability of finding the particle at x_2 is less than half of that of the probability of finding the same at x_3.

From the discussions to this point, we can make some important conclusions:

(i) Schrödinger's equation does not tell us about the position of the particle. Rather it only gives us the probability of finding the particle in the vicinity of a point.

(ii) Statistical descriptions are inherent in quantum mechanics. We can not say much about an electron at a given time. We can only say what happens on average when we make number of observations on the same system. We can also say what happens on average by making many observations simultaneously on similar systems.

(iii) A free particle is that which is not confined to any location. So by letting the value of potential energy $V = 0$ in Schrödinger's equation for a particle in single dimensional case we have its solution in the form as given by equation (1.7) i.e.

$$\psi(x, t) = A\left[\exp j\left(kx - \frac{E}{\hbar}t\right)\right] \tag{1.19}$$

In the expression A can be found by what is called normalization.

But we know that

$$\psi.\psi^* = A\left[\exp j\left(kx - \frac{E}{\hbar}t\right)\right] \times A\left[\exp - j\left(kx - \frac{E}{\hbar}t\right)\right]$$

$$|\psi|^2 = |A|^2 \tag{1.20}$$

Equation (1.20) shows that a free particle can be found with equal probability anywhere, i.e., $|A|^2$ is independent of position. By equation (1.19), it is seen that a free particle does have precisely defined value of energy and hence momentum. But by Heisenberg's Uncertainty Principle, we can define only one of the observables at a time. So, a precisely defined energy or momentum implies an undefined momentum or energy.

A localized free particle must be described as a wave packet formed from superposition of wave functions with different momentum values (hence the k-vectors). This uncertainty in momentum allows spatial localization of the particles.

In single – dimension we can write

$$\psi(x,t) = \int_{-\infty}^{+\infty} A(k)[\exp j(kx - \omega t)]dk \qquad (1.21)$$

In equation (1.21) the wave function in the given integral represents wave packet formed by superimposing wave functions with different momentum (hence k) values.

We have,

$$\lambda = \frac{h}{p} \quad \text{and} \quad k = \frac{2\pi}{\lambda}$$

$$p = \hbar k$$

So, momentum, which is related to the energy of the particle, is also related to k value. Energy E and wave number k are also related to each other.

1.2 Energy Well Model of Metal

1.2.1 Further Application of Schrödinger's Equation

Let an electron be confined to a certain region such that $0<x<L$ where L represents the width of the confinement. Assume that the potential energy of the particle is zero inside the confined region and infinite outside it. The electron thus confined can not leave the region because to leave it requires infinite potential energy.

Let us solve the Schrödinger's equation for the electron confined in an infinite potential well as shown in the figure 1.3.

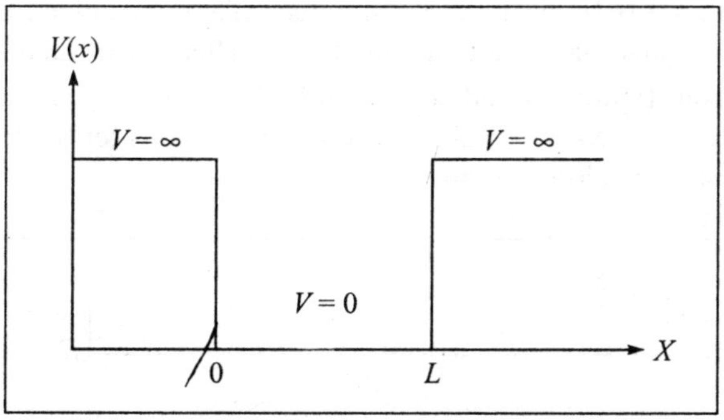

Fig. 1.3 Electron in infinite potential well

A general solution of the Schrödinger's equation (1.16) for single dimensional case is

$$\psi(x) = A \sin \beta x + B \cos \beta x \qquad (1.22)$$

where A and B are some constants that are found from boundary conditions.

And

$$\beta^2 = \frac{8\pi^2 m}{h^2}(E - V)$$

Since the particle can not be found outside the potential well, the wave function representing the particle outside the well will be $\psi = 0$. This condition should be extended to the boundary, i.e. at x=0 and x=L. So, applying boundary conditions, we will get

$$\psi(0) = A \sin 0 + B \cos 0$$

$$\therefore \quad B = 0$$

And,

$$\psi(L) = A \sin \beta L + B \cos \beta L$$

$$A \sin \beta L = 0$$

$$\sin \beta L = \sin(n\pi)$$

$$\beta L = n\pi, \text{ where } n = 1,2,3,\ldots\ldots,\infty.$$

$$\beta^2 = \frac{n^2\pi^2}{L^2}$$

$$\frac{n^2\pi^2}{L^2} = \frac{8\pi^2 mE}{h^2}$$

Here E the total energy of the electron when potential energy is equal to zero (assumed for the electron in infinite potential well) will be given by

$$E_n = \frac{n^2h^2}{8mL^2} \qquad (1.23)$$

$$E_n = \frac{\pi^2 n^2 \hbar^2}{2mL^2} \qquad (1.24)$$

The energy of an electron that is confined in an infinite box (potential well) can not be arbitrary but must be defined by the parameter "n". Alternatively, the energy of an electron confined in an infinite potential well is quantized, i.e., can have only discrete sets of values. These quantized energy values are called "Eigen values" and the corresponding wave functions ψ_n are called Eigen functions.

It can be readily seen from figure 1.4 and equation (1.24) that the energy of a confined particle is quantized, so can have only discrete sets of values. The wave functions for the first three values of n in a single-dimensional case is shown below in figure 1.5.

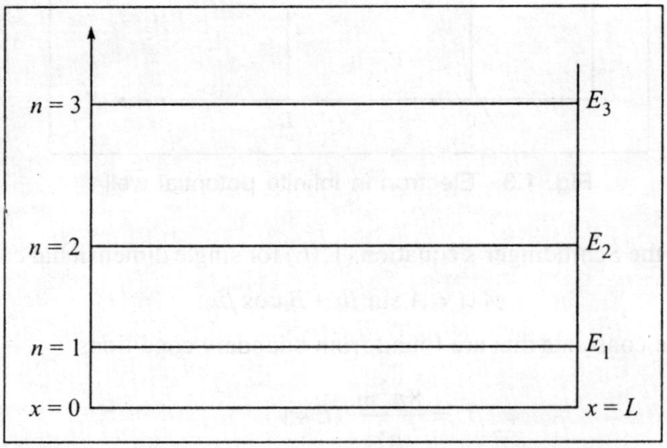

Fig. 1.4 Energy levels of an electron in an infinite potential well

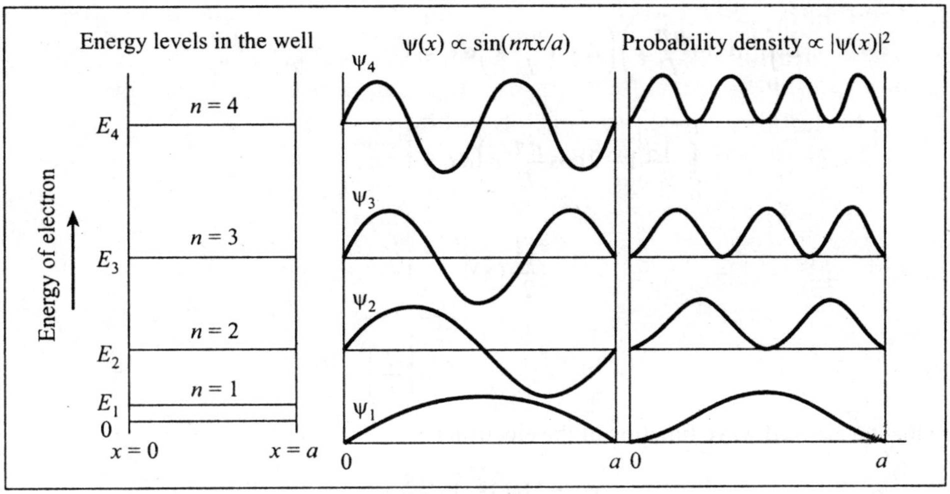

Fig. 1.5 Electron in a single dimensional infinite potential well of width L m. The energy of electron is quantized. The wave functions and probability distribution functions are also shown.

For a particle in infinite potential well the minimum energy that the electron can possess can be calculated by using Heisenberg's uncertainty principle (will be discussed later). Electron in the box at any instant can occupy any position between $x=0$ to $x=L$. At the same time, the electron can be heading to $+x$ direction with a momentum $+p_x$ or to $-x$ direction with momentum $-p_x$, the direction of momentum changing continuously due to collision with the wall of the well. Let Δx and Δp_x be uncertainty in position and uncertainty in momentum of the particle along x axis.

$$\Delta x \, \Delta p_x = L \,.\, 2p_x (\Delta x = L - 0 = L)$$

$$p_x = h/2L$$

Now the energy of the electron consists of only the kinetic one, the potential energy being assumed zero.

$$E = \frac{mv^2}{2} = \frac{p^2}{2m}$$

$$E = \frac{h^2}{8mL^2} \tag{1.25}$$

Now the particle in the infinite potential well can be represented by wave function

$$\psi(x) = A \sin \beta x = A \sin \frac{n\pi}{L} x \tag{1.26}$$

In equation (1.26) A can be determined by normalization. We know that the particle is located within the potential well of width L m. So, we can write

$$\int_0^L \psi . \psi^* \, dx = 1$$

$$L\int_0 \left(A\sin\frac{n\pi}{L}x\right)\left(A\sin\frac{n\pi}{L}x\right)dx = 1$$

$$A^2\int_0^L \frac{1}{2}\left(1-2\cos\frac{n\pi}{L}x\right)dx = 1$$

$$\frac{A^2}{2}(x)\Big|_0^L = 1$$

$$A = \left[\frac{2}{L}\right]^{1/2} \tag{1.27}$$

Hence the normalized wave function of the electron is

$$\psi_n = [2/L]^{\frac{1}{2}}\sin\frac{n\pi}{L}x \tag{1.28}$$

1.2.2 Heisenberg's Uncertainty Principle

The wave function of a free electron corresponds to a traveling wave with a single wavelength. This means the traveling wave extends over all space with the same amplitude. So, the probability distribution function is uniform throughout the whole space. The electron can be found anywhere. This means the uncertainty in position of electron is infinite. At the same time the electron has a fixed well-defined wavelength, which gives the exact value of momentum of the particle. This corresponds to uncertainty in momentum being zero.

Let us recall the case of an electron confined in a single dimensional infinite potential well. The wave function of the particle extends from $x=0$ to $x=L$. So, the uncertainty in position is equal to the width L of the potential well. The momentum of the electron is either $p_x=hk$ in $+x$ direction and $p_x=-hk$ in $-x$ direction. The uncertainty in momentum is therefore difference between momentum in $+x$ direction and $-x$ direction and equal to $2\hbar k$. When the electron is in ground state, $n=1$ and $kL=\pi$. So, the uncertainty in momentum can be written as

$$\Delta p_x = \frac{2\hbar\pi}{L} \tag{1.29}$$

Now, the product of uncertainties in position and momentum of electron can be found.

$$\Delta x\Delta p_x = L\frac{2\pi\hbar}{L}$$

or,

$$\Delta x\Delta p_x = h \tag{1.30}$$

So, the product of uncertainties in position and momentum is equal to Plank's universal constant h. This is one of the fundamental relationships that puts a limit to our knowledge of the behavior of any system. We cannot exactly and simultaneously know or measure both the position and momentum of an electron along a given coordinate.

If Δ_x and Δp_x are the respective uncertainties in position and momentum of an electron during simultaneous measurement along x, then Heisenberg's uncertainty principle state that the their product must be equal or greater than $\hbar$. Mathematically, this relationship can be expressed as

$$\Delta x \Delta p_x \geq \hbar \tag{1.31}$$

Because of the wave nature of quantum mechanics, we can not measure the position and momentum of the electron simultaneously. There will always be uncertainty in position of the electron given by Δx and uncertainty in momentum given by Δp_x. These uncertainties are related by Heisenberg's uncertainty relationship as expressed in (1.31).

These uncertainties are not in any way a consequence of the accuracy of a measurement or the precision of the measuring instrument. Rather, they are theoretical limits to our ability to explore a system and are the part of quantum nature of the universe. If we build the most perfect instrument to measure the position and momentum of the electron at any instant, there will always be uncertainties in momentum and position of the particle such that $\Delta x \Delta p_x \geq \hbar$.

The principle is also used to show for the following:

- the existence of protons and neutrons and non-existence of electrons in nucleus of an atom
- to find the binding energy of electrons in nucleus of an atom
- to find the radius of Bohr's first orbit as well as the finite width of the spectral lines.

1.2.3 Operator Notation

We have, one of the solutions of Schrödinger's equation given in equation (1.19) as

$$\psi(x,t) = A\left[\exp j\left(kx - \frac{E}{\hbar}t\right)\right]$$

Differentiating w.r.t. t, we will get

$$\frac{\partial \psi}{\partial t} = j\frac{E}{\hbar}A\left[\exp j\left(kx - \frac{E}{\hbar}t\right)\right]$$

$$= -j\frac{2\pi E}{h}\psi$$

$$E\psi = j\frac{h}{2\pi}\frac{\partial \psi}{\partial t} \tag{1.32}$$

Thus,

$$E \rightarrow j\frac{h}{2\pi}\frac{\partial}{\partial t} \text{ is called "Energy Operator"}$$

or, $$\hat{E} \rightarrow j\hbar\frac{\partial}{\partial t} \text{ (read energy operator corresponds to)}$$

Similarly, taking partial derivative with respect to x in equation (1.19)

$$\frac{\partial \psi}{\partial x} = jkA\left[\exp j\left(kx - \frac{2\pi E}{\hbar}t\right)\right]$$

$$\frac{\partial \psi}{\partial x} = jk\psi \quad \text{But} \quad k = \frac{2\pi}{h}p = \frac{p}{\hbar}$$

or, $$\frac{\partial \psi}{\partial x} = j\frac{2\pi}{h} p\psi$$

or, $$p\psi = -j\frac{h}{2\pi}\frac{\partial \psi}{\partial x} \tag{1.33}$$

$$\hat{p} \rightarrow -j\frac{h}{2\pi} \text{ is called momentum operator}$$

(Read: momentum operator corresponds to)

Similarly $\hat{x} \rightarrow x$ is called position operator

All these operators are used to predict the respective parameters i.e., energy, momentum and position of the electron.

Let us consider an example, which will help us establish the use of these operators in quantum mechanics and also gives us analogy in statistics so as to understand these operators more clearly.

There are 40 students in the class. And the following is the result of test conducted by the class teacher.

Table 1

Serial No, i	*No of students with score S_i*	*Score S_i*
1	1	20
2	4	19
3	4	18
4	4	17
5	6	16
6	6	15
7	5	14
8	3	12
9	3	11
10	2	10
11	2	08
12	0	00

The average score of the class can be calculated as

$$S_{av} = \frac{\sum\limits_{i=1} N_i S_i}{\sum\limits_{i=1} N_i}$$

$$= \frac{1\times20+4\times19+4\times18+4\times17+6\times16+6\times15+5\times14+3\times12+3\times11+3\times10+2\times8+0\times0}{1+4+4+6+6+5+3+3+2+2+0}$$

$$= 14.935$$

In statistics, S_{av} is called "Expectation value" or "Expected value". In quantum mechanics, we have to find "expected average value" of one of the observables – energy, position or momentum.

The equivalent of S_{av} in quantum mechanics is given by

$$\langle G \rangle = \frac{\int_{-\infty}^{+\infty} \psi^* G \psi dx}{\int_{-\infty}^{+\infty} \psi^* \psi dx} \tag{1.34}$$

In the above equation (1.34), G is one of the operators (e.g. position operator or momentum operator or energy operator).

We have, $$\int_{-\infty}^{+\infty} \psi^* \psi dx = 1$$

Then the equation (1.34) becomes

$$\langle G \rangle = \int_{-\infty}^{+\infty} \psi^* G \psi dx \tag{1.35}$$

1.2.4 Finite Potential Barrier

We have established that the electron confined to an infinite potential well has its energy quantized by virtue of relationship given by equation (1.23) and (1.24). Now let us consider a free electron traveling in $+x$ direction and encountering a barrier whose potential energy is V_0. The energy of the electron E is less than the barrier potential energy also known as "Barrier height". But, unlike in previous case, the potential energy of the barrier is finite.

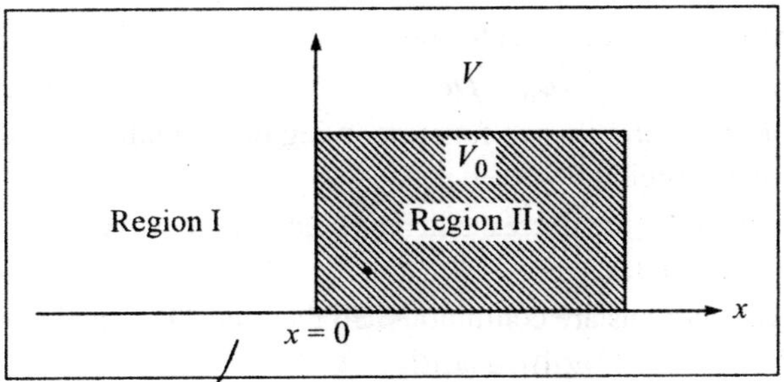

Fig. 1.6 Finite potential barrier

When applying Schrödinger's equation for the electron here, we have to take into account the fact that there are two distinct regions where the behavior of electron will be different.

In region I, electron is free so its potential energy is set equal to zero. So the Schrödinger's equation for the wave function will be

$$\frac{d^2\psi}{dx^2} + \frac{2mE}{\hbar^2}\psi = 0 \tag{1.36}$$

In the region I, $x<0$. For region II where $x>0$, we can write the Schrödinger's equation as

$$\frac{d^2\psi}{dx^2}+\frac{2m}{\hbar^2}(E-V_0)\psi=0 \tag{1.37}$$

The solution of equation (1.36) will be for region I

$$\psi_I = A\exp(j\beta x)+B\exp(-j\beta x) \tag{1.38}$$

For region II,

$$\psi_{II} = Ce^{j\alpha x}+De^{-j\alpha x} \tag{1.39}$$

where $\beta^2=\frac{2mE}{\hbar^2}$ and $\alpha^2=\frac{2m}{\hbar^2}(E-V_0)$

The constants A, B, C and D in the above equations (1.38) and (1.39) can be found by using boundary conditions.

In the region I the electron wave function will be as discussed before but in region II it will be different. α will be imaginary if V_0 is greater than E.

Let $$\gamma = j\alpha$$

Then,

$$\gamma = j\alpha = \left[\frac{2mE}{\hbar^2}(V_0-E)\right]^{\frac{1}{2}} \tag{1.40}$$

The γ given by equation (1.40) is real, not imaginary.

So, the wave function in the region II can now be written as

$$\psi_{II} = Ce^{\gamma x}+De^{-\gamma x} \tag{1.41}$$

When $x\rightarrow\infty$,

$$\psi_{II} = C.\infty + D.0$$

For $\psi.\psi^*$ not to exceed 1, C must be equal to zero.

$$\psi_{II} = De^{-\gamma x} \tag{1.42}$$

The above equation of the electron wave function in region II implies that the amplitude of the wave decreases exponentially in the region.

Note that for larger values of γ, the decrease in amplitude of the wave function in the region II will be faster compared to its smaller values.

At $x = 0$, both the wave functions are continuous. So they must be equal to each other.

$$\psi_I(0) = \psi_{II}(0)$$

or, $$Ae^{j\beta.0}+Be^{-j\beta.0} = De^{-\gamma.0}$$

or, $$A+B = D \tag{1.43}$$

The second boundary condition is: the slope of both the wave functions at x = o are continuous. That means

$$\left.\frac{d\psi_I}{dx}\right|_{x=0} = \left.\frac{d\psi_{II}}{dx}\right|_{x=0}$$

or, $$Aj\beta e^{j\beta x}-Bj\beta e^{-j\beta x} = -\gamma De^{-\gamma x}$$

$$Aj\beta - Bj\beta = -\gamma D \tag{1.44}$$

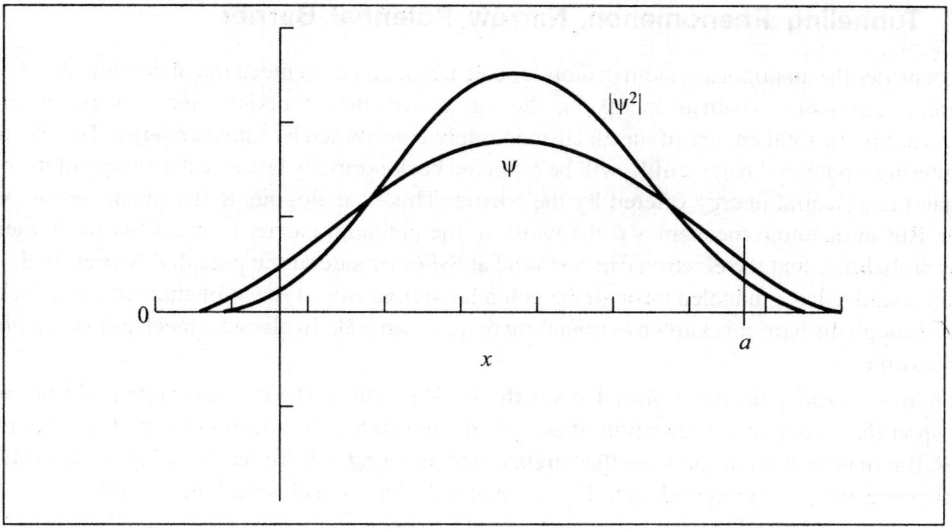

Fig. 1.7 Wave function of electron in the region of finite potential barrier

From equations (1.43) and (1.44) we will find

$$A = \frac{D}{2}\left(1 + j\frac{\gamma}{\beta}\right)$$

$$B = \frac{D}{2}\left(1 - j\frac{\gamma}{\beta}\right) \tag{1.45}$$

So, an electron in any region where it is free or confined can be defined in terms of its amplitude. The same implies to the electron when it is within the finite potential barrier. Depending upon the width of the potential barrier, the amplitude of the electron wave function can be drastically reduced or completely wiped out. This phenomenon can be clearly seen from the figure shown below where the behaviors of wave functions in different regions and for different values of n are shown.

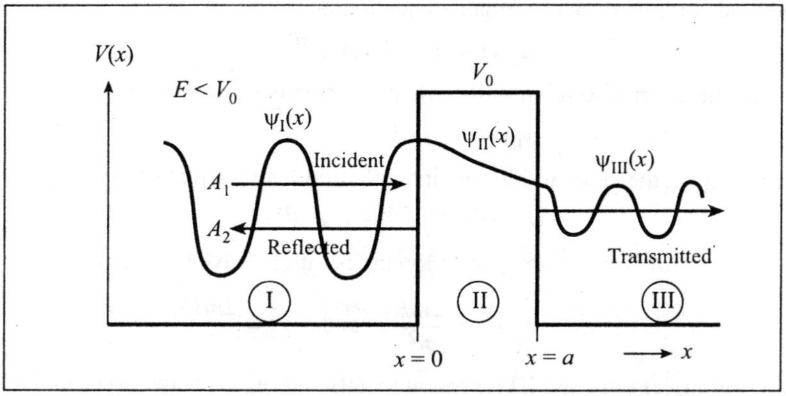

Fig. 1.8 Electron wave functions in different regions.

1.2.5 Tunneling Phenomenon, Narrow Potential Barrier

Let us consider the analogous system in atomic scale i.e. in quantum mechanical domain. As shown in the figure an electron moves with an energy E in the region $x<0$ where for convenience the potential energy is set to zero. So, the total energy of the electron is solely contributed by kinetic energy. The electron when encountering a potential barrier of V_0 will be reflected back normally because the energy of the electron is less than the potential energy offered by the barrier. This is analogous to the phenomenon with roller coaster. But in quantum mechanics if the width of the potential barrier is small (narrow), then there is always probability that the electron can be found at the other side of the potential barrier. In this case the electron is said to have tunneled through the potential barrier. Such type of phenomena, in which electron tunnels through the barrier is known as tunneling or quantum leak. In classical mechanics such phenomena are impossible.

In figure 1.9 a roller coaster is placed at A is shown. When the roller coaster is released from position A, based upon the theory of conservation of energy, it can reach at best to point C and can easily reach to point B. But in no condition can scale the height D and be found at E. To reach point E for the roller coaster the difference energy corresponding to D-A is required. This is in classical mechanics.

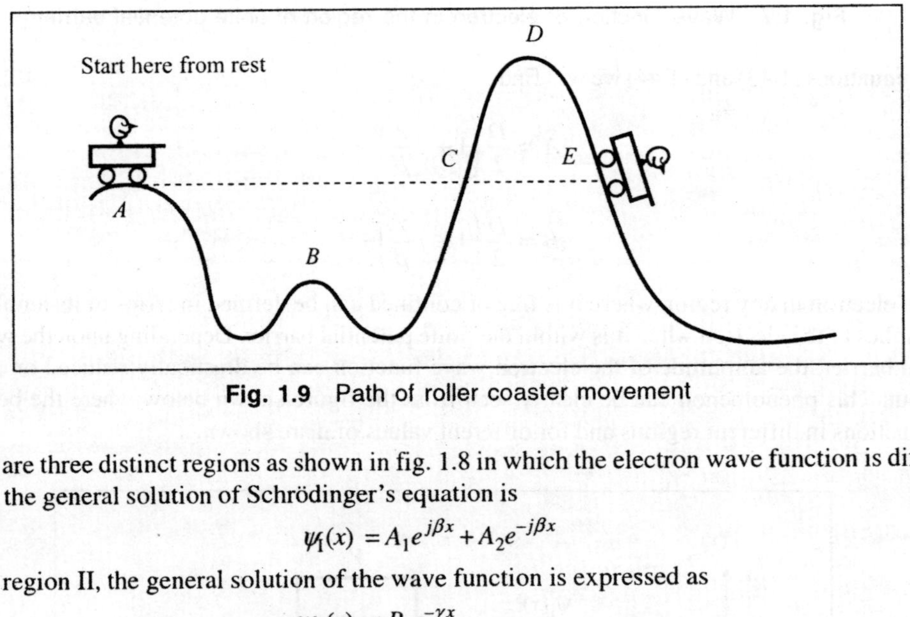

Fig. 1.9 Path of roller coaster movement

There are three distinct regions as shown in fig. 1.8 in which the electron wave function is different. In region I, the general solution of Schrödinger's equation is

$$\psi_I(x) = A_1 e^{j\beta x} + A_2 e^{-j\beta x} \tag{1.46}$$

In the region II, the general solution of the wave function is expressed as

$$\psi_{II}(x) = B_2 e^{-\gamma x} \tag{1.47}$$

Similarly in region III, the general solution of the Schrödinger's equation is

$$\psi_{III}(x) = C_1 e^{j\beta x} + C_2 e^{-j\beta x} \tag{1.48}$$

In equations (1.46), (1.47) and (1.48), the following notations have been used:

$$\beta^2 = \frac{2mE}{\hbar^2} \quad \text{and} \quad \gamma^2 = \frac{2m(V_0 - E)}{\hbar^2}$$

In above description, region I and III represent metals and region II the vacuum. In region I, the wave function is composed of both the incident and reflected waves where as in region III it is only the incident

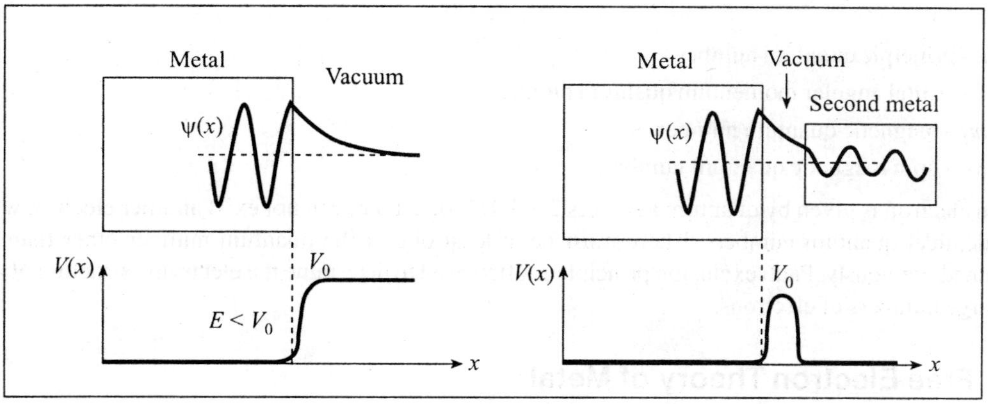

Fig. 1.10 Wave function of an electron incident on potential barrier V_0

wave that gives the wave function. In region II, the wave function decays exponentially given by the expression (1.47) derived earlier.

The relative probability that the electrons will tunnel from region I to the region III through region II is defined as transmission coefficient T which depends very strongly both on the relative potential barrier height $(V_0 - E)$ and the width of the potential barrier. The final expression for the transmission coefficient is expressed as

$$T = \frac{|\psi_{III}(x)|^2}{|\psi_I(incident)|^2} = \frac{C_1^2}{A_1^2} = \frac{1}{1 + D\sinh^2(\gamma L)} \tag{1.49}$$

where
$$D = \frac{V_0^2}{4E(V_0 - E)} \tag{1.50}$$

and γ is the rate of decay of $\psi_{II}(x)$ as expressed above. For a wide or high barrier, using $\gamma L \geq 1$ in equation (1.49) and $\sinh(\gamma L) \approx \frac{1}{2}\exp(-2\gamma L)$, we can deduce

$$T = T_0 \exp(-2\gamma L) \tag{1.51}$$

The expression (1.51) gives the probability of tunneling through the potential barrier of width L and height V where T_0 is given by

$$T_0 = \frac{16E(V_0 - E)}{V_0^2} \tag{1.52}$$

The reflection coefficient R is given by

$$R = \frac{A_2^2}{A_1^2} = 1 - T \tag{1.53}$$

1.2.6 Pauli Exclusion Principle

This principle based on experimental observations states that no two electrons within a given system may have all four identical quantum numbers n, l, m_l and m_s. Each set of values for n, l, m_l and m_s represent a possible electronic state and correspondingly a wave function $\psi_{n,\,l,\,m_l,\,m_s}$

where

n - principle quantum number

l – orbital angular momentum quantum number

m_l – magnetic quantum number

m_s – spin magnetic quantum number.

If an electron is given by quantum numbers 2,1,1,1/2, then there can not exist another electron with all four identical quantum numbers. There must be at least one of the quantum number other than those mentioned previously. Pauli exclusion principle is often used to determine the electronic structure of atoms with large numbers of electrons.

1.3 Free Electron Theory of Metal

1.3.1 General Concepts

To account in qualitative and quantitative way for the characterization of metallic properties, a theory was precluded. Drude (1900) constructed a theory of metal on the basis of highly successful kinetic theory of gases considering as a gas of electrons. But in gas one type of particles are present and in metal at least two, i.e., electrons and ions are present. In this theory interaction of electrons with ions and other electrons is rejected. Electrons are assumed to achieve thermal equilibrium with surroundings through collisions which are instantaneous events.

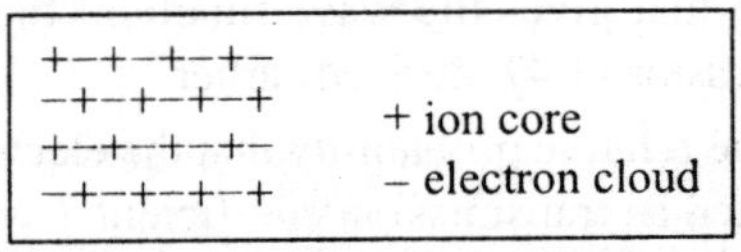

Fig. 1.11 Ion core and electron cloud

According to this theory, the outermost electrons known as valence electrons of an atom in metal are very loosely attached to the parent atom, so they are not bound to the parent atom. These electrons are free to move throughout the whole metal. These electrons are attached to different atoms at different instants. These freely roaming electrons within the metal form what is known as "Electron gas". When an atom loses its valence electron it becomes positively charged ion. So, the metallic bond is an unsaturated covalent bond due to electrostatic attraction between the positively charged ion core and negatively charged electron cloud.

The cohesion in metallic crystal is due to the positive ion core and negative electrons passing between these ions. This is the reason for metallic bonds being weaker than ionic and covalent bonds, which are called saturated bonds. This indicates to the ductile nature of metal.

In free electron theory it is assumed that the potential due to the ion cores is uniform throughout the metal. This means a free electron has the same potential energy everywhere within the metal. Although, strictly saying, the potential energy of an electron due to electrostatic attraction in the metal has a finite negative value, we assume it zero because we are concerned with change in potential energy rather than the potential energy itself.

So, if the electron has kinetic energy only

$$E = \frac{mv^2}{2} \tag{1.54}$$

But

$$\lambda = \frac{h}{p} = \frac{h}{mv}$$

$$k = \frac{2\pi}{\lambda} \quad \text{and} \quad v = \frac{\hbar k}{m}$$

Substituting for λ and k in equation (1.54), we will have

$$E = \frac{m}{2}\left(\frac{\hbar^2 k^2}{m^2}\right)$$

$$E = \frac{\hbar^2 k^2}{2m} \tag{1.55}$$

The equation (1.55) represents the kinetic energy of the free electron within the metal.

The relationship between E and k given by equation (1.55) is parabolic in nature as shown in figure 1.12.

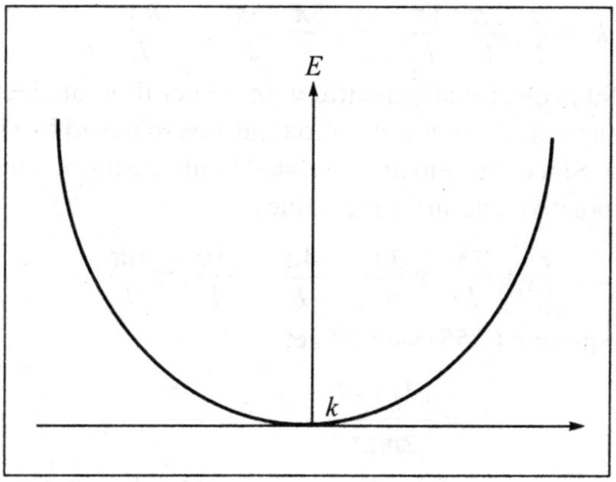

Fig. 1.12 E-k diagram of free electron in metal

1.3.2 Electron in a Linear Solid (Metal)

Consider a copper wire of length L. There are many free electrons within that linear solid. The wavelength exhibited by electron in the wire will depend upon their kinetic energy.

For minimum kinetic energy the wavelength of electron will be greatest and will progressively go down as the kinetic energy increases. So, for minimum kinetic energy

$$L = \frac{\lambda}{2} \tag{1.56}$$

And for higher values of kinetic energies, the relationship between wavelength and length of the wire is given by

$$L = 2 \times \frac{\lambda}{2}$$

$$L = 3 \times \frac{\lambda}{2}$$

And so on up to n

$$L = n \times \frac{\lambda}{2}$$

Now, arranging for wavelength in terms of length of the wire, we will get

$$\gamma = 2L, L \frac{2}{3} L, \frac{L}{2}, \frac{2}{5} L, \frac{1}{3} L, \ldots \frac{2}{n} L \tag{1.57}$$

Wavelength and wave number are related as

$$k = \frac{2\pi}{\lambda}$$

Substituting the value of wavelength from equation (1.57) in the above expression, we will get

$$k = \frac{2\pi}{2L}, \frac{2\pi}{L}, \frac{2\pi}{\frac{2}{3}L}, \frac{2\pi}{L/2}, \frac{2\pi}{2L/5}, \frac{2\pi}{L/3}, \ldots, \frac{2\pi}{2L/n}$$

$$k = \frac{\pi}{L}, \frac{2\pi}{L}, \frac{3\pi}{L}, \frac{4\pi}{L}, \frac{5\pi}{L}, \frac{6\pi}{L}, \ldots, \frac{n\pi}{L} \tag{1.58}$$

In normal condition, a solid is electrically neutral with no net flow of electrons in any direction. This means a free electron having a velocity in one direction must be matched by similar electron having same velocity in opposite direction. Since the velocity is related to momentum, which in turn is related to wave number must have both the positive and negative values.

$$k = \pm\frac{\pi}{L}, \pm\frac{2\pi}{L}, \pm\frac{3\pi}{L}, \pm\frac{4\pi}{L}, \pm\frac{5\pi}{L}, \pm\frac{6\pi}{L}, \ldots, \pm\frac{n\pi}{L} \tag{1.59}$$

Now substituting for k in equation (1.55) we will get

$$E = \frac{n^2\hbar^2\pi^2}{2mL^2} \tag{1.60}$$

Above expression (1.60) shows that a free electron in a metal does not possess any arbitrary value of energy rather its energy is quantized by virtue of E being function of principle quantum number n.

1.3.3 Fermi Energy

In metals if to say in very few words Fermi Energy is defined as the highest filled energy level at a temperature of absolute zero, i.e., 0°K. All the energy levels up to the Fermi level are filled at 0°K and empty above it. In E-k diagram Fermi energy can be shown as in figure 1.13. The distribution of electrons among various energy levels in accordance with Pauli's exclusion principle. If N is the total number of electrons to be accommodated on the line, then for even n, we have $N = 2n_F$, where n_F is principle quantum number of the Fermi level. Thus, for $n=n_F$ from equation (1.60), we can rewrite the expression as

$$E_F = \frac{\hbar^2}{2m}\left(\frac{\pi n_F}{2}\right)^2 \tag{1.61}$$

Thus, the Fermi energy depends upon the length of the box and the number of electrons in the box.

$$E_F = \frac{\hbar^2}{2m}\left(\frac{\pi N}{2L}\right)^2 = \frac{\hbar^2}{2m}k_F^2 \tag{1.62}$$

where k_F is the Fermi wave vector and the sphere of radius k_F in E-K space is called Fermi sphere. The number of occupied electrons in the sphere is given by

$$N = 2\times\frac{\text{Volume of sphere of radius } k_F}{\text{Volume of element of } k\text{ - space}} = \frac{2\frac{4}{3}\pi k_F^3}{(K_x)^3} = \frac{L^3 k_F^3}{3\pi^2}$$

or,

$$N = \frac{Vk_F^3}{3\pi^2} \qquad \text{where} \quad K_x = \frac{2\pi}{L}$$

And,

$$n = \frac{N}{V} = \frac{k_F^3}{3\pi^3}$$

In the above expression, 2 comes from the two allowed values of the spin quantum number.

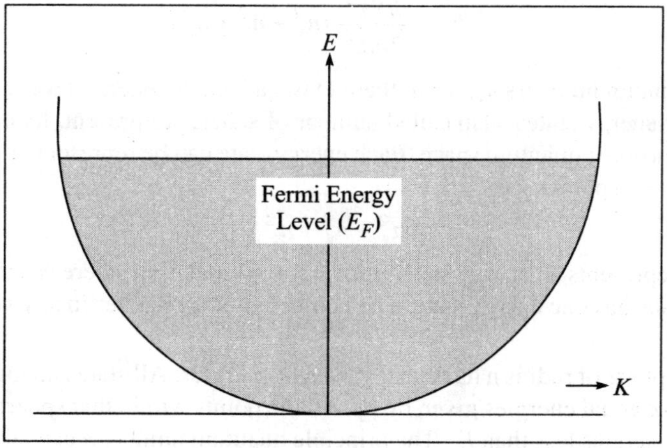

Fig. 1.13 The highest filled energy level at 0°K

1.3.4 Degenerate State

In a single dimensional solid the energy of the electron is expressed as

$$E = \frac{\pi^2 n^2 \hbar^2}{2mL^2} = \frac{n^2 h^2}{8mL^2} \tag{1.63}$$

This expression is valid both for free electron and electron in a potential well.

In a three dimensional solid which is the general case, the above expression will be supplemented by the fact that in place of principle quantum number n is taken for single dimension and would be replaced by corresponding three-dimensional value.

$$n^2 = n_x^2 + n_y^2 + n_z^2 \tag{1.64}$$

And the energy expression becomes

$$E_n = \frac{\pi^2 \hbar^2}{2mL^2}(n_x^2 + n_y^2 + n_z^2) \tag{1.65}$$

where, as in the earlier cases, n_x, n_y and n_z can only be integers, i.e., 1,2,3,4,.......,∞.

The set of n values along different axes in three-dimensional case will determine the energy associated with the electron. Unlike in single dimension, the same n along x axis can represent different energies because of the fact that its energy is not only dependent upon principle quantum number along x-axis but also upon n along *y* and *z* axes also. When different combinations of principle quantum number represent same energy, these are called "**degenerate states**". For example $n(1,1,2)$, $n(1,2,1)$ and $n(2,1,1)$ will represent the same electron energy. These all three electronic states are degenerate states.

1.4 Density of States Function

1.4.1 Definitions of Number of States, Density of States

The quantized energy state of electron is expressed as

$$E_n = \frac{\hbar^2\pi^2}{2mL^2}(n_x^2 + n_y^2 + n_z^2) \tag{1.66}$$

For each set of quantum numbers n_x, n_y, n_z there exists a specific energy level called energy state. To determine how many energy states, also called number of states, are present, let us consider a quantum number in three dimensional quantum space. Each energy state can be represented by a vector n which is the radius from origin to a point n_x, n_y, n_z.

$$n^2 = n_x^2 + n_y^2 + n_z^2 \tag{1.67}$$

Every integer (n) represents an energy state. For n_x=1, n_y=1 and n_z=1, there is only one energy state. In other words, a unit cube has one energy state. The number of energy states in any volume is proportional to that volume.

On the surface of sphere of radius n has equal values of energies. All states laying on the surface of the sphere of radius *n* have equal energies given by E_n. All the points within that sphere of radius *n* represent quantum states with energies less than E_n. The principle quantum number *n* can take integer values only, so the energy states are defined only in positive octant of the sphere made by the radius vector *n*.

So, the number of energy states within the sphere of radius n and having energies less than E_n is given by

$$N = \frac{1}{8}\frac{4\pi n^3}{3} \tag{1.68}$$

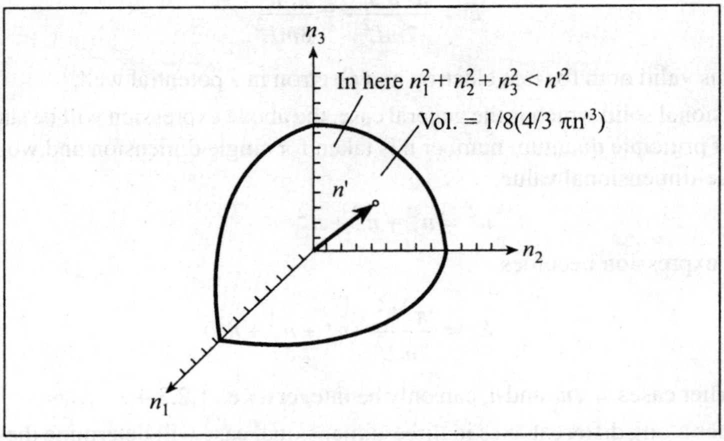

Fig. 1.14 Sphere created by radius vector *n*

But we have, from equation (1.57)

$$n^2 = \frac{2mL^2}{\hbar^2} E = \frac{8mL^2}{h^2} E$$

$$n = \left[\frac{8mL^2}{h^2} E\right] \tag{1.69}$$

Substituting for n in equation (1.58) yields

$$N = \frac{1}{8}\frac{4\pi}{3}\left[\frac{8mL^2}{h^2} E\right]^{3/2}$$

$$N = \frac{\pi}{6} K^{3/2} E^{3/2} \tag{1.70}$$

where $$K = \frac{8mL^2}{h^2} \tag{1.71}$$

The expression (1.70) gives us the number of energy states lying within the positive octant of the sphere of radius n. We have taken only $1/8^{th}$ of the volume of sphere, which is its volume of positive octant. The number of states represents the number of energy levels in quantum space.

Differentiating equation (1.70) with respect to energy E yields

$$\frac{dN}{dE} = \frac{\pi}{4} K^{3/2} E^{1/2} \tag{1.72}$$

In the above expression, $\frac{dN}{dE} = Z(E) =$ Density of states

For an electron placed in three-dimensional potential box with all sides equal to L, then the volume of the box is

$$V = L^3$$

Then the density of states function becomes

$$Z(E) = \frac{dN}{dE} = \frac{\pi}{4}\left[\frac{2^2 \times 2mL^2}{h^2}\right]^{3/2} E^{1/2}$$

$$= \frac{\pi}{4}\frac{8L^3}{h^3}(2m)^{3/2} E^{1/2}$$

$$= 2\pi V \frac{(2m)^{3/2}}{h^3} E^{1/2}$$

$$= CE^{1/2}$$

where $$C = 2\pi V \frac{(2m)^{3/2}}{h^3} \tag{1.73}$$

$$Z(E) = CE^{1/2} \tag{1.74}$$

Equations (1.72) and (1.74) represent the density of states function. The density of states function can be defined as "number of energy states per unit energy in an energy interval of *dE*".

In the above figure (1.15) the curve representing $Z(E)$ is hyperbola where all energy levels less than E_n are taken it to account. The shaded area represents the number of energy states in an elementary energy interval *dE*.

$$dN = Z(E)dE \tag{1.75}$$

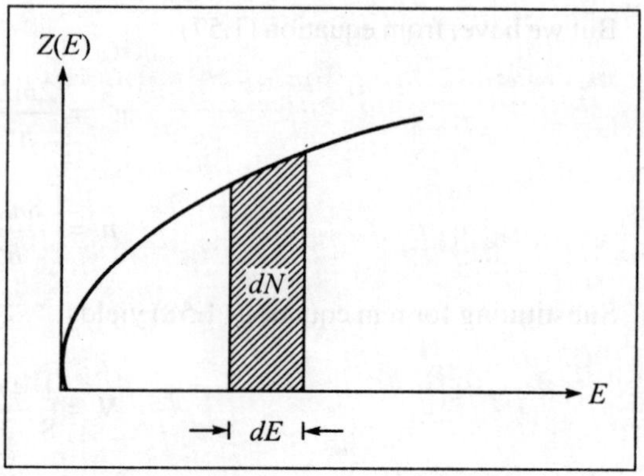

Fig. 1.15 Density of states vs. energy functions

From figure (1.15) and equation (1.75) it is clear that higher the energy, the more are the number of states available and lower the energy, fewer are the energy states available. In quantum mechanics we are concerned not only about the number of states and density of states, but also about how much of these energy states are actually occupied by electrons. We never can say precisely how many of these states are or can be occupied by electrons but can with fair accuracy predict this. To predict the number of energy states occupied by electrons, we will use suitable statistical tools that are applicable in quantum mechanics.

1.5 Fermi-Dirac Distribution Function

To describe the classical physics Maxwell-Boltzmann statistics is used. But this is not applicable for systems in quantum mechanical domain. In quantum mechanics the distribution functions are governed by the fact that similar indistinguishable particle can be differentiated according to their spatial location and orientation. To describe the energy distribution in quantum mechanical system, we use what is called Fermi-Dirac distribution function. The particles, which obey the Fermi-Dirac distribution function, are called fermions. Fermions obey the Pauli exclusion principle and only one fermion can occupy a given quantum state at once.

Fermi-Dirac distribution function is written as

$$F(E) = \frac{1}{\left(1 + e^{\frac{E-E_F}{kT}}\right)} \tag{1.76}$$

where $F(E)$ = occupation index or probability of occupation

E = Energy of the electron

E_F = Fermi energy level

T = absolute temperature

$k = 1.38 \times 10^{-23}$ J/K, called Boltzmann's universal constant.

Figure (1.16) shows the Fermi-Dirac distribution function for three different temperatures.

From expression (1.76), when $E<E_F$, at T=0K

$$F(E) = \frac{1}{1 + \exp\left(-\frac{\Delta E}{kT}\right)} = \frac{1}{1 + e^{-\infty}} = 1$$

where $\Delta E = E_F - E.$

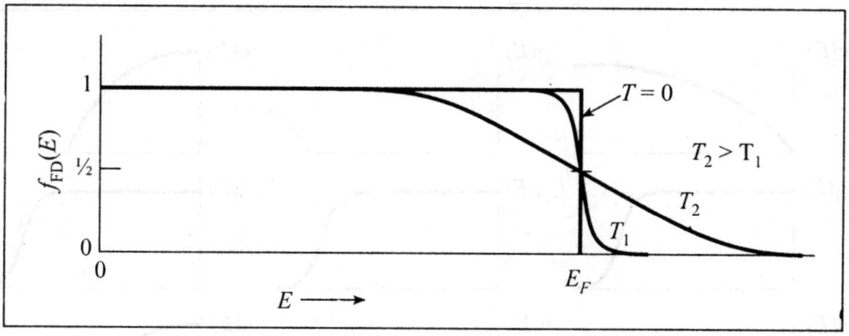

Fig. 1.16 Fermi-Dirac distribution

When at T=0K, E=E_F, the Fermi-Dirac distribution function becomes indeterminable and its value changes from 1 to 0 for the same condition.

For $$T = 0\text{K and } E > E_F$$

$$F(E) = \frac{1}{1 + e\infty} = 0$$

i.e. the probability of occupation of energy state with energy greater than Fermi energy at absolute zero is zero. Simply saying, electron cannot occupy a state higher than Fermi level at 0K.

The higher the temperature, the higher is the probability of higher energy states being occupied by the electron. One interesting property of the Fermi energy is that for any temperature the occupation index is 0.5 if the energy of the electron is equal to Fermi energy. This can be simply understood as the higher temperature imparts energy to the electron in the metal , then the electrons can be excited to higher energy levels, so the probability of finding them in these regions increases.

At higher temperatures, the Fermi-Dirac distribution function approximates to Boltzmann's classical distribution function. The occupation index rapidly decreases from 1 to 0 within the energy interval $2\Delta E$.

When $E >> E_F$,

$$F(E) \approx \exp\left[-\frac{E - E_F}{kT}\right] \tag{1.77}$$

By using the Fermi function or the occupation index we can calculate the number of energy states actually occupied by the electrons. Number of energy states per unit energy in an energy interval dE is simply found by multiplying the density of states function $Z(E)$ by probability of occupation function $F(E)$. But to account for Pauli exclusion principle, which states that each state is occupied by two electrons with opposite spins, we should multiply again by 2. So, mathematically, we can write

$$N(E) = 2Z(E)F(E) \tag{1.78}$$

or,

$$N(E) = 2CE^{1/2} \frac{1}{1 + \exp\left[\frac{E - E_F}{kT}\right]} \tag{1.79}$$

In above expression (1.79), $N(E)$ is called "Population density "of electrons and gives account for number of electrons per unit energy in the given energy interval.

The three functions density of states function: $Z(E)$, probability of occupation function $F(E)$ and population density function $N(E)$ are plotted below in figure 1.17.

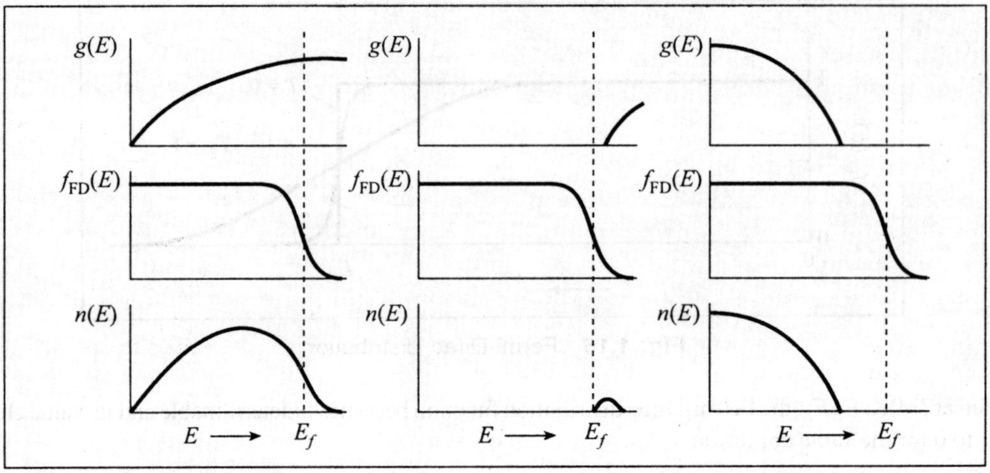

Fig. 1.17 Three different functions - $Z(E)$, $F(E)$ and the corresponding $N(E)$.

In the above figure 1.17 the area under the curve in $N(E)$ vs E represents the number of electrons with energy equal to or less than E_n. For an elementary energy interval of dE the number of electrons can be found as

$$dN = N(E)dE \tag{1.80}$$

Integrating equation (1.80) from $E = 0$ to $E = E_F$, we can get

$$N' = \int_0^{E_F} N(E)d(E)$$

$$= \int_0^{E_F} 2CE^{1/2}dE$$

$$= \frac{4}{3}CE_F^{3/2}$$

$$= \frac{4}{3}\frac{2\pi V}{h^3}(2m)^{3/2}E_F^{3/2}$$

$$= \frac{8\pi V}{3h^3}(2m)^{3/2}E_F^{3/2}$$

Now, let n be the number of electrons per unit volume, then

$$n = \frac{N'}{V} = \frac{8\pi}{3h^3}(2m)^{3/2}E_F^{3/2} \tag{1.81}$$

Solving equation (1.81) for E_F yields

$$E_F = \frac{h^2}{2m}\left(\frac{3n}{8\pi}\right)^{\frac{2}{3}} \tag{1.82}$$

By using expression (1.82), Fermi Level for any metal can be calculated if number if electrons per unit volume for that metal is known. And, conversely, if Fermi level is given we can calculate the electron concentration, i.e., number of electrons per unit volume.

Putting $h = 6.624 \times 10^{-34}$J.s

$m = 9.1 \times 10^{-31}$kg in the expression (1.82), we obtain

$$E_F = 3.65 \times 10^{-19} \times N^{2/3}\ \text{eV} \tag{1.83}$$

where $1\text{eV} = 1.609 \times 10^{-19}$J

1.6 Thermionic Emission, Work Function

When metal is heated scores of electrons within it become more energetic and at some point they can even leave the surface of the metal. The process of ejection of electrons from the surface of the metal at rather high temperature is called ***thermionic emission***. Not all the electrons, which are excited by heating, are able to leave the metal surface. There is a threshold (minimum) value of energy, which the electron must possess if it is to leave the metal surface. This threshold value of energy is called ***work function*** of the metal. Work function is denoted by Greek alphabet Φ.

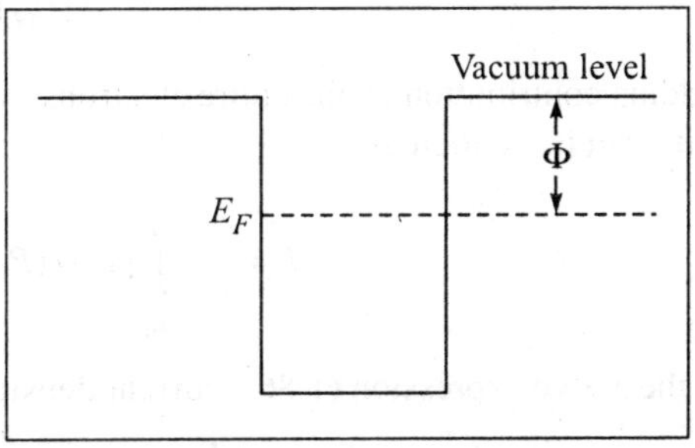

Fig. 1.18 Fermi level and work function in metal

At T=0K, all the states up to Fermi level are filled and above are empty. An electron needs at least an energy equal to $E_F + \Phi$ to escape from the metal. We assume that the potential energy of the electron is zero and it only possesses kinetic energy. Then, we can express its energy in terms of momentum in three-dimensional geometry as

$$E = \frac{1}{2m}(P_x^2 + P_y^2 + P_z^2) \tag{1.84}$$

For the electron to be able to escape the metal it must have in addition to right value of energy, right direction as well. For, example if x is the coordinate perpendicular to metal surface, the electron momentum must satisfy the following condition

$$\frac{P_x^2}{2m} \rangle \frac{P_{x0}^2}{2m} = E_F + \Phi \tag{1.85}$$

where Px_0 is the threshold momentum.

But by satisfying the above condition does not mean that the electron can escape the metal. In quantum mechanics, there is still probability that the electron will be reflected from the barrier. So, we must introduce a new parameter known as reflection coefficient $r(P_x)$.

$$\text{Probability of escape} = 1 - r(P_x)$$

The number of electrons per unit volume having momentum between P_x and $P_x + dP_x$ is given by $N(P_x)dP_x$ in analogy with the number of electrons in energy interval between E and $E + dE$ which was given by $N(E)dE$.

Multiplying $N(P_x)dP_x$ by velocity of electrons associated with the momentum P_x (thus the kinetic energy $\frac{P_x^2}{2m}$), we will get the number of electrons arriving at the surface of metal per unit second per unit time given by $\frac{P_x}{m}N(P_x)dP_x$.

$$\left[\because \frac{d(p_x^2/2m)}{dP_x} = \frac{P_x}{m}\right]$$

Similarly, number of escaping electrons is given by

$$(1 - r(P_x))\frac{P_x}{m}N(P_x)dP_x$$

Adding contribution by the entire electrons, which have momenta greater than P_{x0}, the emission current density can be written as

$$J = \frac{e}{m}\int_{P_{x0}}^{\infty}[1 - r(P_x)]P_xN(P_x)dP_x, \text{ A/m}^2 \tag{1.86}$$

In the above expression (1.86), current density

$$J = \frac{I}{A}, \quad \text{and} \quad I = \frac{dQ}{dt} = \frac{de}{dt}$$

where I = current in A,
A = surface area, m^2
Q = charge
e = electron charge
t = time interval

The number of electrons in an incremental momentum range can be obtained in the manner similar to the number of electrons in incremental energy range.

Comparing equations (1.65) and (1.84), we obtain

$$P_x = \frac{h}{2}n_x, \quad P_y = \frac{h}{2}n_y \quad \text{and} \quad P_z = \frac{h}{2}n_z \tag{1.87}$$

We know that the number of states in a cube of side 1 is equal to 1. Similarly, the number of states in volume of sides dn_x, dn_y, dn_z is given by

$$\left[\frac{2}{h}\right]^3 dP_x dP_y dP_z \tag{1.88}$$

As in the case of number of states with momentum not exceeding the given P_{x0}, P_{y0} and P_{z0} is found by multiplying equation (1.88) by 1/8.

$$\frac{1}{8}\left[\frac{2}{h}\right]^3 dP_x dP_y dP_z \tag{1.89}$$

$$= \frac{dP_x dP_y dP_z}{h^3}$$

Now, as we found density of states function in an energy interval, we can find the same in respective momentum interval and then by multiplying equation (1.89) by 2 to account for counter spinning electrons

$$Z(P_x, P_y, P_z) = 2\frac{dP_x dP_y dP_z}{h^3}\frac{1}{dP_x dP_y dP_z} \tag{1.90}$$

$$Z(P_x, P_y, P_z) = \frac{2}{h^3}$$

The number of electrons in momentum range $P_x + dP_x$, $Py + dP_y$ and $P_z + dP_z$ is given by

$$N(P_x, P_y, P_z)\, dP_x dP_y dP_z = \frac{2}{h^3}\frac{dP_x dP_y dP_z}{1 + e^{\frac{E - E_F}{kT}}} \tag{1.91}$$

where $$E = \frac{1}{2m}(P_x^2 + P_y^2 + P_z^2)$$

To determine the number of electrons in the momentum range P_x to $P_x + dP_x$, the expression (1.91) should be integrated for all values of P_y and P_z.

$$N(P_x)dP_x = \frac{2}{h^3}\int_{-\infty}^{\infty}\int_{-\infty}^{\infty}\frac{dP_y dP_z}{\exp\left[\left\{\frac{1}{2m}(P_x^2 + P_y^2 + P_z^2) - E_F\right\}/kT\right] + 1} \tag{1.92}$$

When the work function $\Phi >> kT$, the unity term from the denominator of expression (1.92) can be neglected. And, by using integral table, we get

$$N(P_x)dP_x = \frac{4\pi mkT}{h^3}\exp(E_F/kT)\exp(-P_x^2/2mkT)dP_x \tag{1.93}$$

By comparing equations (1.83) and (1.93) and taking $r(P_x) = r$, we get

$$J = A_0(1 - r)T^2 e^{-\frac{\Phi}{kT}} \text{ - Richardson's equation} \tag{1.94}$$

where $$A_0 = \frac{4\pi mk^2}{h^3} = 1.2 \times 10^6 \text{ Am}^{-2}\text{K}^{-2} \tag{1.95}$$

In expression (1.94), the emission current density is heavily dependent upon both the work function of the material and its temperature. Since both work function and temperature figure up in exponential function, a small change in either of these two parameters will cause the emission current density to change by very large amount. The increase in temperature and decrease in work function both will cause the emission current density to increase exponentially.

The stages in measuring the thermionic emission current are discussed below.

1.6.1 Schottky Effect

As a new approach for measuring thermionic emission we will use first a method known as image charge method in electrostatics. Based upon image charge method, any charge placed at a distance x from a conducting surface can be substituted by the original charge at the initial position and an image charge with opposite polarity that is situated at a distance similar to that of original charge but in just opposite direction.

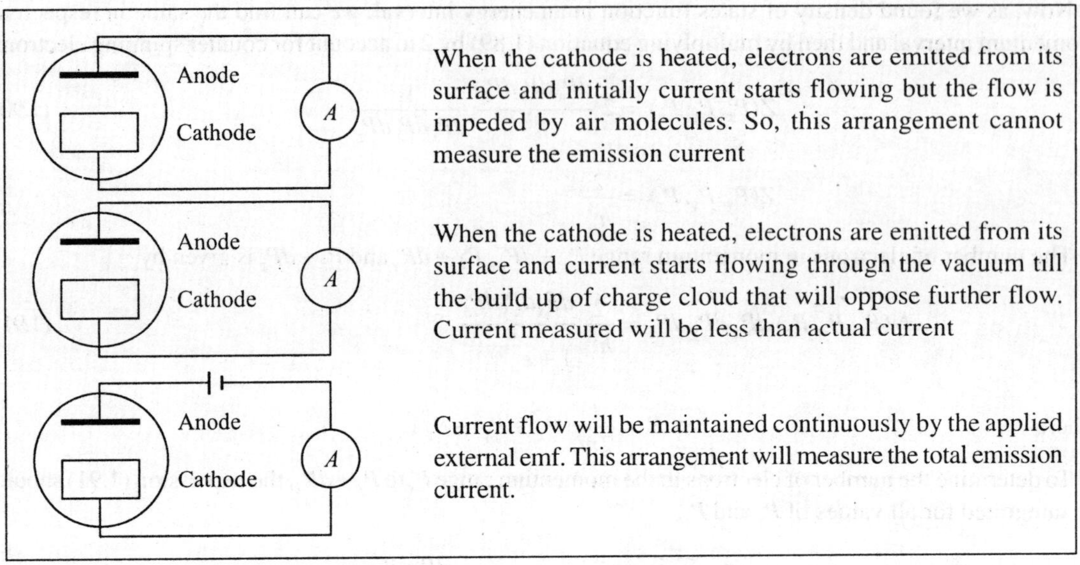

Fig. 1.19 Measurement of emission current (a) in open air, (b) in vacuum tube (c) in vacuum tube with the application of external emf.

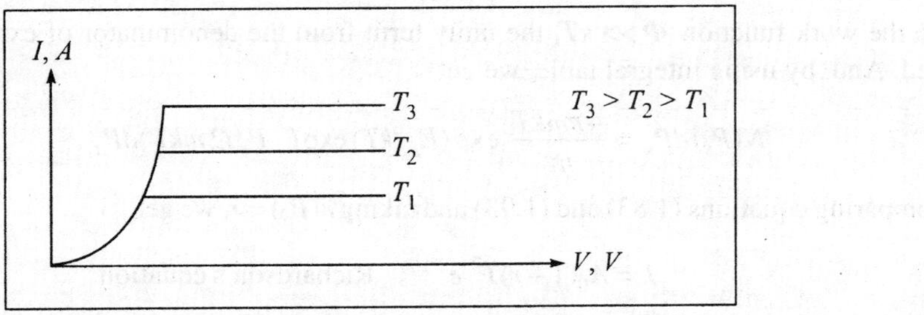

Fig. 1.20 Emission current as a function of applied voltage for different temperatures.

The force of electrostatic attraction between the real and image charge can be calculated by Coulumb's law.

$$F = \frac{e^2}{4\pi\varepsilon_0 (2x)^2} = \frac{e^2}{16\pi\varepsilon_0 x^2} \tag{1.96}$$

where e = charge of an electron = 1.6×10^{-19}C

ε_0 = absolute permittivity of the air = 8.854×10^{-12} F/m

The potential energy can be found be as an integral of force F given by equation (1.96) from x to ∞.

$$V(x) = \int_x^{\infty} F(x)dx = \frac{e^2}{16\pi\varepsilon_0 x^2} \tag{1.97}$$

Now, if we include this image field, then the potential barrier at metal-vacuum interface becomes as shown in figure (1.22).

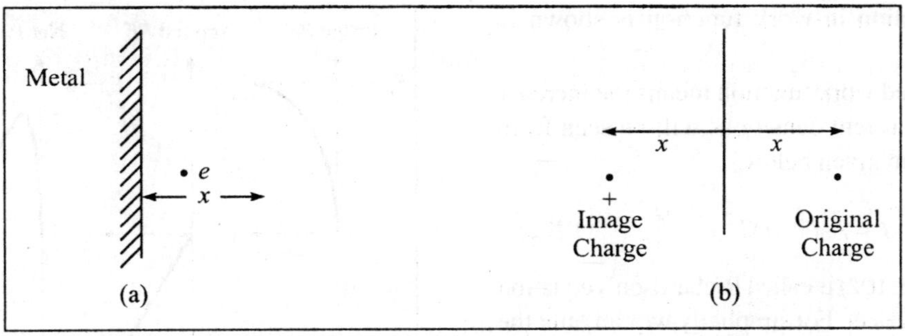

Fig. 1.21 Charge metal interface (a), with image charge (b)

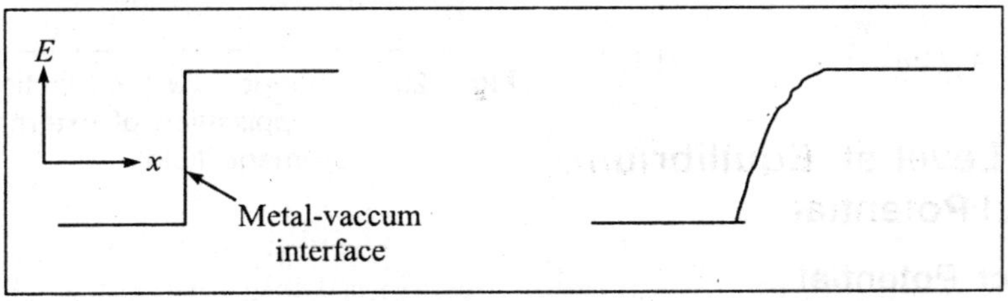

Fig. 1.22 (a) without image charge, (b) with image charge

The application of image field does not itself make any difference in the work function and potential energy barrier for electron at metal-vacuum interface. But if apply external electric field, and then there will be clear difference.

If E is the applied electric field, then potential energy due to this field will be

$$V(x) = -eEx \tag{1.98}$$

Now, by adding the contributions from image field and applied external field, the potential energy will be

$$PE(x) = E_F + \Phi - \frac{e^2}{16\pi x} - eEx \tag{1.99}$$

The reduction in work function or barrier height will be maximum if

$$\frac{d}{dx}\left(-\frac{e^2}{16\pi\varepsilon_0 x} - eEx\right) = 0$$

$$V_{max}(x) = -\left(\frac{e^3 E}{16\pi\varepsilon_0}\right)^{1/2} \tag{1.100}$$

$V_{max}(x)$ is the maximum reduction in potential energy barrier after taking into account at the effect of image field and applied external field.

The effective work function can now be written as

$$\Phi_{eff} = \Phi - \left(\frac{e^3 E}{16\pi\varepsilon_0}\right)^{1/2} \tag{1.101}$$

The reduction in work function is shown in figure 1.23.

The reduced work function means the increase in emission current density as will be seen from the expression given below.

$$J = A_0(1-r)T^2 e^{-\frac{\Phi_{eff}}{kT}} \quad (1.102)$$

Equation (1.102) is called Richardson's equation for Schottky effect. For simplicity we can omit the reflection coefficient taking its value equal to zero and the equation (1.102) can be written as

$$J = A_0 T^2 e^{-\frac{\Phi_{eff}}{kT}} \quad (1.103)$$

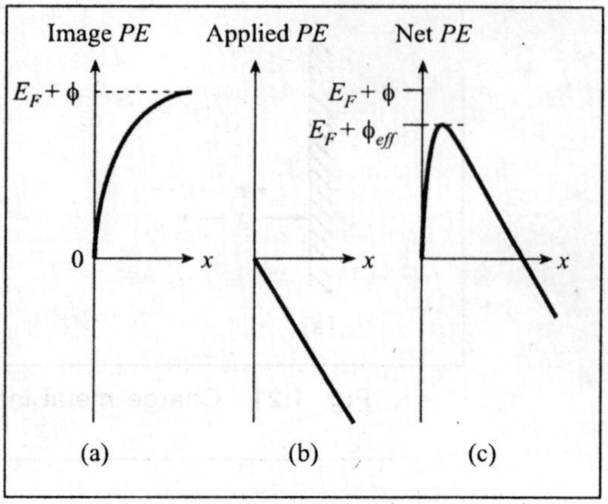

Fig. 1.23 Change in work function as the result of application of external field and use of image field.

1.7 Fermi Level at Equilibrium, Contact Potential

1.7.1 Contact Potential

As we already know each metal has different work function. For each metal different values of energy are required so that electron can be emitted. Two metals with different work functions are brought together.

Each metal has different work function signifying that their Fermi levels are different. But when these two metal with different Fermi level (work function as well) are brought into contact, electrons from metal 1 with higher Fermi level will start crossing over to metal 2 with lower Fermi level by tunneling to occupy the lower energy states that are vacant. This electron transfer from metal 1 to metal 2 reduces the total energy of the electrons in metal1-metal2 system. Metal 1 having lost electrons becomes positively charged whereas metal2 having received electrons becomes negatively charged. Consequently a potential difference is developed at the junction called ***contact potential***. This process continues till the contact potential is large enough to prevent the further transfer of electrons from metal 1 to 2. So the system reaches equilibrium. At equilibrium, the Fermi level of both the metal will be same. The ***contact potential*** due to difference in work functions of the metals brought into contact can be calculated as

$$e\Delta V = \Phi_2 - \Phi_1$$

$$\Delta V = \frac{\Phi_2 - \Phi_1}{e} \quad (1.104)$$

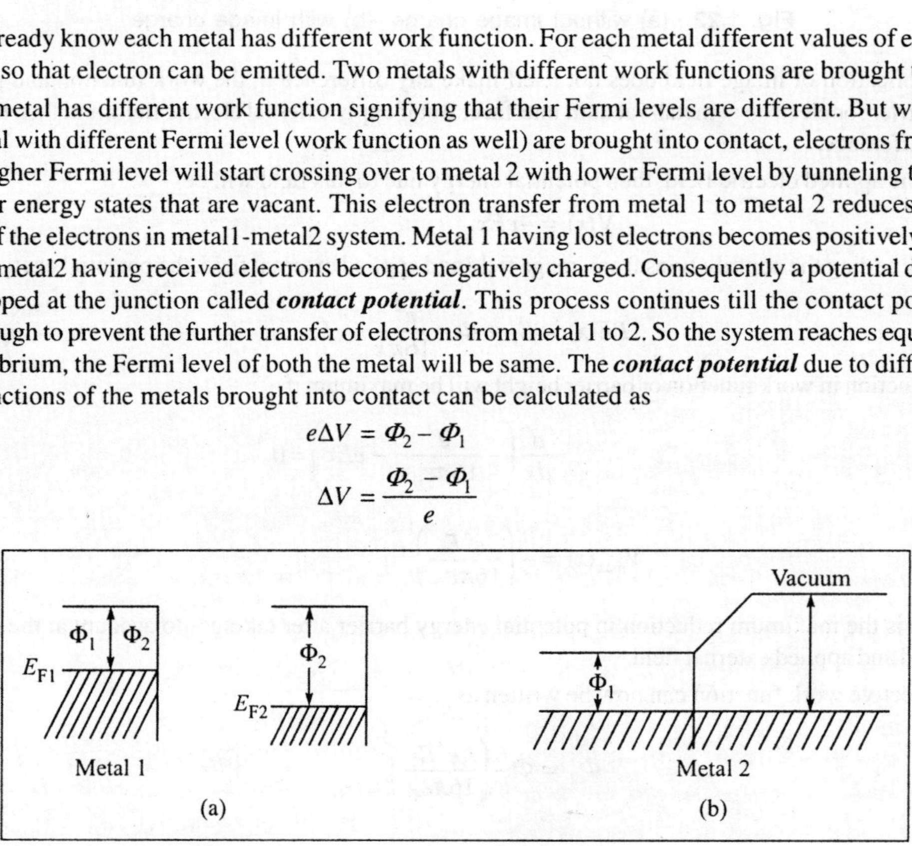

Fig. 1.24 (a) when metals are separate, (b) when metals are brought together.

1.7.2 Seebeck Effect

When two metals, one hot and the other cold are brought together, more energetic electrons in the hot metal will diffuse to the cold metal where electrons are less energetic. The electrons diffusing to cold side from the hot leave behind exposed positive ions whereas in the hot region there is excess of negative charge. So, there is net potential difference between the hot metal and cold metal due to the difference in temperature. This is called ***Seebeck effect***, which states that "thermo-emf in a thermocouple can be set maintaining temperature gradient without using any cell or generator".

$$S = \frac{\Delta V}{\Delta T} \tag{1.105}$$

where S = Seebeck coefficient or thermo-electric power

ΔV = induced potential difference or thermo-emf

ΔT = temperature difference

If V_{HC} is the thermo-emf of the thermocouple arrangement, then we can write

$$V_{HC} \propto \Delta T \quad \text{i.e.} \quad V_{HC} = S\Delta T \tag{1.106}$$

The values of Seebeck coefficient for different metals are different. S represents potential of cold junction with respect to the hot junction. Mostly S is negative but Cu has S positive. According to Mott_Jones, the Seebeck coefficient is given by

$$S = -\frac{\pi^2 k^2 T}{3eE_{F0}} \tag{1.107}$$

where E_{F0} is the Fermi energy at 0K.

Seebeck effect can be utilized to measure temperature in thermocouple arrangement where one junction (normally cold junction) is maintained at a constant reference temperature and the temperature of the hot junction is measured. The potential at each junction will depend upon their respective Seebeck coefficients. And if we connect these two junctions by wire and put a voltmeter across them, we can measure the thermoelectric potential difference. If ΔV_H and ΔV_C are the potentials of hot and cold junctions and S_H and S_C are their respective Seebeck coefficients, the potential difference can be found as

$$V_{HC} = \Delta V_H - \Delta V_C = \int_{T_0}^{T} (S_H - S_C)\, dT \tag{1.108}$$

where T_0 and T are the temperatures of cold junction (reference junction) and hot junction respectively.

For any thermo-couple, we can relate the potential difference and temperature difference as

$$V_{HC} = a\Delta T + b(\Delta T)^2 \tag{1.109}$$

Equation (1.109) is called thermocouple equation in which a and b are constants. The plot of V_{HC} and ΔT is shown in figure 1.25, where T_N is called neutral temperature at which thermo-emf is maximum. The temperature T_i at which the thermo-emf again falls to zero is called inversion temperature. For Fe-Cu thermo-couple, T_N= 270°C and T_i= 540°C.

1.8 Solved Numerical Examples

1. Calculate the wavelength of a 50 gram golf ball traveling at a velocity of 20ms^{-1}.

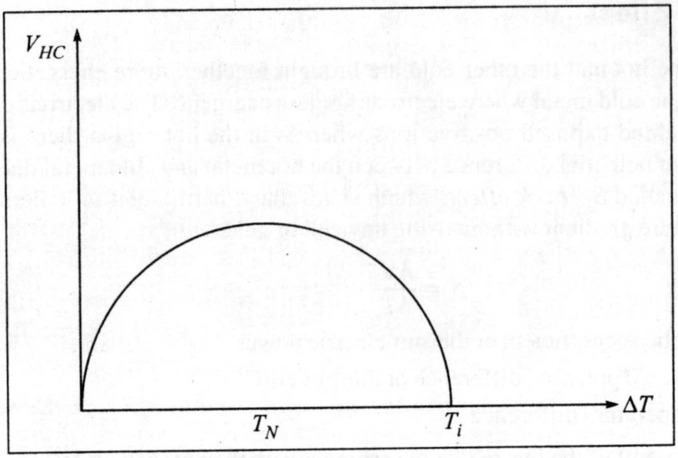

Fig. 1.25 Variation of thermo-emf with change in temperature

Solution:

We have, according to de Broglie's equation, wavelength of the golf ball

$$\lambda = \frac{h}{p} = \frac{h}{mv} = \frac{6.624 \times 10^{-34}\ \text{Js}}{(50 \times 10^{-3}\ \text{kg})\,(20\,\text{ms}^{-1})} = 6.63 \times 10^{-34}\ \text{m}$$

The wavelength is so small that the golf ball will not exhibit any wave effect.

2. Calculate the wavelength of a proton traveling at 2200ms^{-1}

Solution:

Mass of the proton = 1.67×10^{-27}kg

So, the wavelength can be found from

$$\lambda = \frac{h}{p} = \frac{h}{mv} = \frac{6.624 \times 10^{-34}\ \text{Js}}{(1.67 \times 10^{-27}\ \text{kg})\,(2200\text{ms}^{-1})} = 1.8 \times 10^{-10}\text{m} = 0.18\ \text{nm}$$

3. Find the wavelength of an electron accelerated by 100V.

Solution:

The whole potential energy (=eV) will be converted in to kinetic.

$$KE = PE$$

$$\frac{p^2}{2m} = \text{eV}$$

$$p = (2\text{meV})^{\frac{1}{2}}$$

Now the wavelength of the electron will be

$$\lambda = \frac{h}{p} = \frac{h}{(2\text{meV})^{\frac{1}{2}}} = \frac{6.624 \times 10^{-34}\ \text{Js}}{[(9.11 \times 10^{-31}\text{kg})\,(1.6 \times 10^{-19}\,C)\,(100\,V)]^{1/2}}$$

$$= 1.23 \times 10^{-10}\text{m} = 0.123\text{nm}$$

4. What is the energy of blue photon with a wavelength of 450nm?

Solution:

The energy of the photon is given by

$$E = hf \quad \text{and} \quad f = \frac{c}{\lambda} \quad \text{where } c = 3\times 10^8 \text{m/s}$$

$$E = \frac{(6.624\times 10^{-34}\text{ Js})(3\times 10^8\text{ ms}^{-1})}{450\times 10^{-9}\text{ m}} = 4.4\times 10^{-19} J$$

which is very small and is generally expressed in electron-volt.

So, the electron energy $E = 2.75$ eV.

5. In the photoelectric experiment, green light, with a wavelength of 522 nm is the longest wavelength radiation that can cause photoemission of electron from a clean sodium surface. Calculate the work function of sodium. If ultraviolet radiation with a wavelength 250 nm is incident to the sodium surface, what will be the kinetic energy of the photo-emitted electrons?

Solution:

At threshold, the photon energy just causes the photoemission, that is, the electrons just overcome the potential barrier Φ. Thus the work function of sodium

$$hf = \Phi$$

$$\Phi = \frac{hc}{\lambda} = \frac{(6.624\times 10^{-34}\text{ Js})(3\times 10^8\text{ ms}^{-1})}{522\times 10^{-9}\text{ m}} = 2.38\text{eV}$$

The energy of the incoming photon is

$$E_1 = \frac{hc}{\lambda_1} = \frac{(6.624\times 10^{-34}\text{ Js})(3\times 10^8\text{ ms}^{-1})}{250\times 10^{-9}\text{ m}} = 4.96\text{eV}$$

Only 2.38eV is required to eject electrons from the surface of sodium. So, the excess energy (4.96eV-2.38eV)=2.58eV will go to the electron in the form of kinetic energy.

6. An electron is confined to an infinite potential well of size 0.1nm. Calculate the ground energy of the electron and radian frequency. How this electron can be put to the third energy level?

Solution:

The energy of electron confined to an infinite potential well is given by

$$E_n = \frac{n^2 h^2}{8mL^2}$$

$n = 1$ for ground state and $L = 0.1$nm. So,

$$E_1 = \frac{1\times(6.624\times 10^{-34}\text{ Js})^2}{8\times(9.11\times 10^{-31}\text{ kg})(0.1\times 10^{-9})^2} = 37.6\text{eV}$$

Radian frequency of electron associated with this energy is

$$\omega = \frac{E}{\hbar} = \frac{37.6\times 1.6\times 10^{-19}\text{ J}}{1.055\times 10^{-34}\text{ Js}} = 5.71\times 10^{16}\text{ rad.s}^{-1}$$

Energy of electron in third energy level is

$$E_3 = n^2E_1 = 3^2 \times 37.6\text{eV} = 338.4\text{eV}$$

The short fall in energy required to put the electron in to the third level is

$$E_3 - E_1 = 338.4\text{eV} - 37.6\text{eV} = 300.8\text{eV}$$

This difference or short fall can be provided by a photon having exactly the same energy, no less, no more. The wavelength of such photon is given by

$$\lambda = \frac{hc}{E} = \frac{(6.624 \times 10^{-34}\ \text{Js})\,(3 \times 10^{8}\ \text{ms}^{-1})}{300.8 \times 1.6 \times 10^{-19}\ \text{J}} = 4.12\,\text{nm}$$

Photon with a wavelength of 4.12 nm is a x-ray photon.

7. An apple of 100g is confined to move within a rigid crate of 1m size. What is the minimum speed of the object? What should be the quantum number *n* if the object is moving with a speed of 1m/s?

Solution:

Since the object is within rigid walls, we can *PE* outside the walls as an infinite and calculate energy as

$$E_n = \frac{n^2h^2}{8mL^2}$$ This energy will be minimum for n=1. So,

$$E_1 = \frac{1^2(6.624 \times 10^{-34}\ \text{Js})^2}{8 \times (0.1\,\text{kg})\,(1\,\text{m})^2} = 5.45 \times 10^{-67}\,J = 3.4 \times 10^{-48}\ \text{eV}$$

So, the minimum velocity can be calculated using the above kinetic energy

$$E_1 = \frac{mv^2}{2}$$

$$v = (2E_1/\text{m})^{1/2} = (2 \times 5.45 \times 10^{-67}\ \text{J}/0.1\ \text{kg})^{1/2} = 3.3 \times 10^{-33}\ \text{ms}^{-1}$$

This speed can not be measured by any instrument and thus for all practical purposes the apple within the crate is at rest.

8. For an electron confined to an infinite potential well of width 0.1 nm, determine the uncertainty in momentum and kinetic energy.

Solution:

Heisenberg's uncertainty principle is expressed as

$$\Delta p_x \Delta x \geq \hbar$$

$$\Delta p_x \approx \frac{\hbar}{\Delta x} = \frac{1.054 \times 10^{-34}\ \text{Js}}{0.1 \times 10^{-9}\ \text{m}} = 1.054 \times 10^{-24}\ \text{kg.m.s}^{-1}$$

Now, the uncertainty in kinetic energy associated with this uncertainty in momentum

$$\Delta E(KE) = \frac{\Delta p_x^2}{2m} = \frac{(1.054 \times 10^{-24}\ \text{kg.m.s}^{-1})^2}{2 \times 9.11 \times 10^{-31}\ \text{kg}} = 6.1 \times 10^{-19}\,\text{J} = 3.81\text{eV}$$

9. Estimate the minimum velocity of an apple of 100g confined to a crate of 1 m size.

Solution:

Uncertainty in momentum of apple is

$$\Delta p_x \approx \frac{\hbar}{m\Delta x}$$

$$\Delta v_x = \frac{\Delta p_x}{m} = \frac{\hbar}{m\Delta x} = \frac{1.054 \times 10^{-34}\ \text{Js}}{(0.1\,\text{kg})\,(1\,\text{m})} = 1.054 \times 10^{-33}\ \text{ms}^{-1}$$

Quantum nature of the universe implies that the apple in the crate is moving with a velocity of the order of 10^{-33}ms^{-1} and will take the apple to cover an atomic distance of 0.1 nm around 10^{19} years.

10. Prove the uncertainty principle $\Delta p_x\, \Delta x \geq h$ from an analogy with the classical wave frequency f, having an uncertainty of frequency Δf when measured over a finite period of Δt.

Solution:

Let f_1 and f_2 be two frequencies which differ by Δf when measured over Δt. Time to observe one beat is $1/\Delta f$ second. One beat may certainly be observed if $\Delta t 1/\Delta f$ which will lead to $\Delta t\, \Delta f \geq 1$

The distance traveled by wave in time Δt will be $\Delta x = v\, \Delta t$

$$\Delta x \geq v/\Delta f$$

But $$f = v/\lambda \text{ or, } \frac{\Delta f}{\Delta \lambda} = -\frac{v}{\lambda^2}$$

or $$\frac{v}{\Delta f} = \frac{\lambda^2}{\Delta \lambda}$$

$\therefore \quad \Delta x \Delta \lambda \geq \lambda^2$ If λ is measured over distance Δx, the wavelength is uncertain by $\Delta \lambda$. We also have

$$\lambda = \frac{h}{p}$$

$$\left|\frac{\Delta \lambda}{\Delta p}\right| = \frac{h}{p^2}$$

$$\Delta \lambda = \frac{h}{p^2}\Delta p$$

$$\Delta x \frac{h}{p^2} \Delta p \geq \lambda^2$$

So, $$\Delta x \frac{h}{p^2} \Delta p \geq \frac{h}{p^2}$$

$$\Delta x \Delta p \geq h$$

We can come to another interesting and important conclusion. Since energy is associated with frequency, the frequency uncertainty will lead to energy uncertainty.

$$\Delta E = h\Delta f$$

But, $$\Delta f \Delta t \geq 1$$

$$\Delta f \geq \frac{1}{\Delta t}$$

So, $$\Delta E \geq \frac{h}{\Delta t}$$

$$\Delta E\, \Delta t \geq h$$

A particle having an energy E for an interval of Δt will have its energy uncertain by ΔE.

11. Calculate the number of photons emitted by a radio station broadcasting at 792 kHz operating at 5kW.

Solution:

$$\text{Photon energy} = hf = (6.624 \times 10^{-34}\ \text{Js})(792 \times 10^3 \text{s}^{-1}) = 5.246 \times 10^{-25}\ \text{J}$$

$$\text{Number of photon/sec} = 5.246 \times 10^{-25}\text{J}/(5 \times 1000\ \text{Js}^{-1}) = 9.53 \times 10^{27}\text{s}^{-1}$$

12. Evaluate the probability that an energy state kT above the Fermi level will be occupied by an electron.

Solution:

Here $$E = E_F + kT$$

$$F(E) = \frac{1}{1 + \exp[(E - E_F)/kT]} = \frac{1}{1 + \exp[(E_F + kT - E_F)/kT]}$$

$$= \frac{1}{1 + e} \approx 0.27$$

13. A transmitter type vacuum tube has a cylindrical Th-coated W cathode, which is 4 cm long and 2mm in diameter. Estimate the saturation current if the tube is operated at 1600°C. The emission constant A_0 = 3×10^4A m^{-2} K^{-2} for thorium on tungsten. Work function for Th on W is 2.6 eV.

Solution:

Temperature, T = 1600 + 273 K = 1873 K.

Richardson's equation is

$$J = A_0 T^2 \exp\left[-\frac{\phi}{kT}\right] = (3 \times 10^4\ \text{Am}^{-2}\text{K}^{-2})(1873\ \text{K})^2 \exp\left[-\frac{2.6 \times 1.6 \times 10^{-19}}{(1.38 \times 10^{-23}\ \text{JK}^{-1})(1873\ \text{K})}\right]$$

$$J = 1.08 \times 10^4\ \text{Am}^{-2}$$

The emission surface area is

$$A = \pi dL = \pi \times 2 \times 10^{-4}\ \text{m} \times 4 \times 10^{-2}\ \text{m} = 2.5 \times 10^{-4}\ \text{m}^2$$

The saturation current or the maximum current obtainable is

$$I = JA = (1.08 \times 10^4\ \text{Am}^{-2})(2.5 \times 10^{-4}\ \text{m}^2) = 2.7\ \text{A}$$

14. Calculate the saturation emission current for the example 8, if the cathode is separated from anode by 1mm and the anode voltage is 4kV.

Solution:

The field at the cathode is

$$E = V/d = 4 \times 1000\text{V}/(1 \times 10^{-3}\text{m}) = 4 \times 10^6\ \text{V/m}$$

The effective work function is

$$\Phi_{\text{eff}} = \Phi - e(eE/16\Phi\varepsilon_0)^{1/2}$$

$$\Phi_{eff} = 2.6 - (1.6\times 10^{-19}\,\text{C})\times\frac{1}{2}\times\left(\frac{(1.6\times 10^{-19}\,\text{C})\,(4\times 10^{6}\,\text{V/m}}{4\pi\times 8.854\times 10^{-12}\,\text{F/m}}\right) = 2.52\text{eV}$$

The new saturation current is

$$I_{new} = A_0T^2\exp(-\Phi_{eff}/kT) = (3\times 10^4\,\text{Am}^{-2}\,\text{K}^{-2})\,(1873\,\text{K})^2\exp\left[-\frac{2.52\times 1.6\times 10^{-19}\,\text{J}}{1.38\times 10^{-23}\,\text{JK}^{-1}\times 1873\,\text{K}}\right]$$

$$= 4.3\text{A}$$

Exercise

1. Calculate the De Broglie wave-length associated with a proton moving with a velocity equal to $\frac{1}{20}$ th of the velocity of light. [2.6×10^{-14} m]
2. Calculate the smallest possible uncertainty in the position of an electron moving with velocity $3\times 10^7\text{ms}^{-1}$. [$3.8\times 10^{-12}$ m]
3. An excited state of H-atom has life time of 2.0×10^{-14}s. Calculate the minimum error with which the energy of the state can be measured. [0.033eV]
4. Calculate the kinetic energy of a neutron having de Broglie wave-length 1Å. [0.082eV]
5. If an electron is confined a box of length 10^{+2} Å, calculate the minimum uncertainty in its velocity. [$1.1\times 10^4\text{ms}^{-1}$]
6. Find the energy of an electron moving in one dimension in an infinitely high potential box of width 1.5 Å. [16.735eV]
7. Prove that the density of states of free electrons (N) and energy E is $Z(E) = \frac{3N}{2E}$.
8. Show that the average kinetic energy per electron for 3D free electron gas at 0K is $\frac{3}{5}E_{F0}$.
9. The density of potassium is 860 Kgm^{-3}. Calculate the Fermi energy at 0K. [2.02eV]
10. X-rays of wavelength 0.91Å fall on a metal plate having work function 2eV. Find the wavelength associated with emitted photoelectrons.
11. Calculate the density of states for 1m^3 of Na at Fermi level for T=OK and for 1g-mole of Na.
12. Discuss the significance of wave-particle duality in quantum mechanics.
13. What is the reason behind some materials being heavier than others? Discuss.
14. Explain why silver is a better conductor compared to iron.

Free Electron Theory of Conduction in Metals

Chapter 2

2.1 Crystalline Structure

A crystalline solid is a solid in which atoms bond with each other in a regular pattern to form a periodic collection or array of atoms. All crystals have periodicity, which leads to long-range order. Local bonding geometry is repeated many times at regular intervals to produce periodic array of atoms that constitute crystal. The location of each atom is well known by virtue of periodicity. Nearly all metals, many ceramics, semiconductors and various polymers are crystalline in solid. The periodic array in which the atoms are located is called the crystal lattice. The arrangement of space positions of the atoms in a crystal is called space lattice or lattice array.

Since a crystal is essentially a periodic repetition of small volume of atoms in three dimensions, it is useful to identify the repeating unit so that the crystal properties can be described through this unit. The most convenient small cell in the crystal structure that carries the properties of the crystal is called unit cell. The repetition of unit cell in three dimensions generates the whole crystal.

There are different types of unit cell geometries of crystals of different materials. In cubic type geometries, there are basically three types of unit cells, namely Simple cubic, Face centered cubic and Body centered cubic. Not all the elements and compounds have cubic structures; there are other unit cell geometries like orthorhombic, monoclinic, hexagonal, tetragonal, etc. In cubic structures, there are diamond cubic structure and zinc blende cubic structure.

2.1.1 Simple Cubic Structure

Atoms are placed at the corners of the unit cell geometry and by repeating this in all directions crystal is formed. The length of a side is equal to the distance between two atoms. Each atom touches along the cube edges only. The structure is loosely packed.

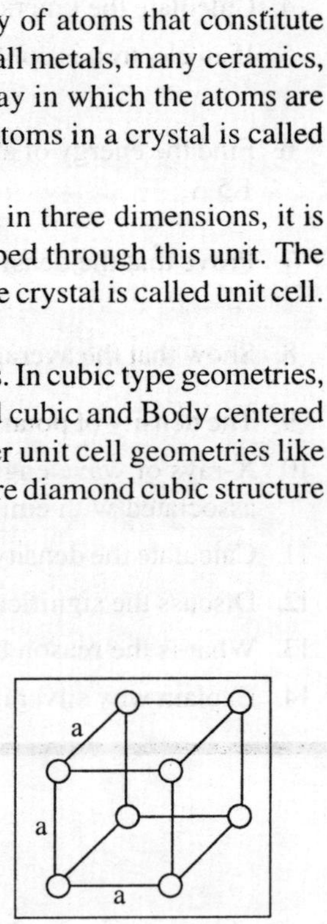

Fig. 2.1 Unit cell of SCC

Only $1/8^{th}$ of each of the atoms belongs to a unit cell. So, each unit cell has $8 \times 1/8 = 1$ atom. The packing density is defined as the ration of volume actually occupied by an atom to the volume of the unit cell.

$$\text{Packing Density} = \frac{V_{attom}}{V_{unit-cell}} \tag{2.1}$$

So, for SCC we have

$$\text{Packing Density} = \frac{\frac{4}{3}\pi R^3}{a^3} = \frac{\frac{4}{3}\pi R^3}{(2R)^3} \approx 0.52$$

2.1.2 Face Centered Cubic (FCC) Structure

Here the atoms are placed as in the case of SCC at all the corners of the unit cell and additional atoms are placed at the intersection of face diagonals in all faces. Half of the atom at each of the face diagonals belongs to the unit cell. So the total number of atoms in a unit cell equals to (1 + 6*1/2 = 4) four. Each FCC unit cell has four atoms in it. The number of neighboring atoms is termed as coordination number. In FCC structure the coordination number is 12.

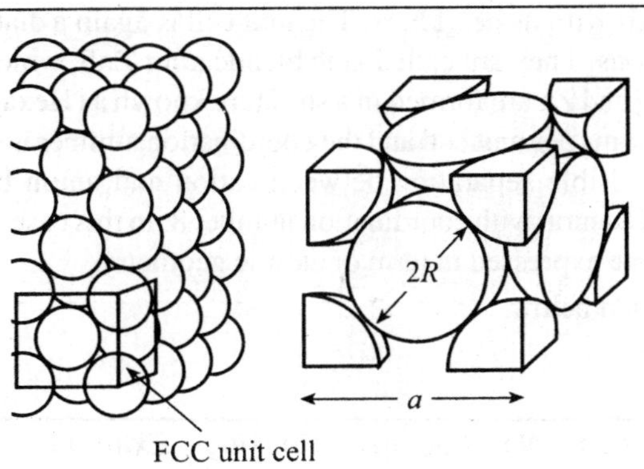

$$\text{Face diagonal} = 4R$$

$$\text{Side } a = 4R/\sqrt{2}$$

$$\text{Body diagonal} = [(4R)^2 + (4R/\sqrt{2})^2]^{1/2} = 2\sqrt{6}R$$

$$PD = \frac{4 \times (4/3) \times \pi \times R^3}{(2\sqrt{2} \times R^3)}$$

Fig. 2.2 FCC unit cell geometry.

2.1.3 Body Centered Cubic (BCC) Structure

In this type of unit cell structure in addition to atoms at the corner of the cell geometry, there is an atom occupying the central place within the unit cell. All the corner atoms are touching the central atom. The coordination number here is 8 and the number of atoms per unit cell = 1 + 1 = 2. Elements with this type of cell geometry have fairly high packing density.

In BCC structure, there is an atom at the center of the unit cell.

$$\text{Body Diagonal} = R + 2R + R$$

$$\text{Face Diagonal} = \frac{2BD}{\sqrt{3}} = \frac{4\sqrt{2R}}{\sqrt{3}}$$

$$\text{Side } a = \frac{FD}{\sqrt{2}} = \frac{4R}{\sqrt{3}}$$

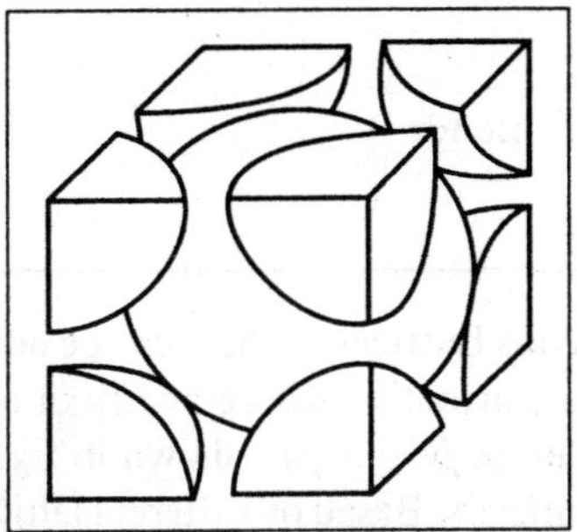

Fig. 2.3 BCC unit cell geometry

$$\text{Packing Desity} = \frac{\text{Volume Occupied by atoms}}{\text{Volume of the unit cell}}$$

$$= \frac{2 \times \frac{4\pi R^3}{3}}{\left(\frac{4R}{\sqrt{3}}\right)^3} = \frac{8\pi R^3}{3} \times \frac{3\sqrt{3}}{64R^3}$$

$$= \sqrt{3}\frac{\pi}{8} \approx 0.68$$

Covalently bonded solids such as Si and Ge have diamond cubic structure. Here the neighboring atoms bond with each other in tetrahedral configuration. In additions to the atoms at the corner and intersections of the face diagonals there are four atoms within the unit cell. So there are altogether 8 atoms /unit cell. In GaAs crystal each atom forms four directional bonds with its neighbors. The unit cell is again a diamond type cubic with Ga and As atoms alternating positions. They are called zinc blende after ZnS, which has this type of unit cell. Atoms of elements like Co, Mg and Zn are formed in a structure known as Hexagonal Close Packed (HCP) structure in which there are 2 atoms per unit cell and the coordination number is 12. In CsCl structure, which is BCC structure, there is visible separation between cation and anion bound together. Each CsCl unit cell consists of 1 cation and 1 anion with coordination number 8. In this case, since there are different atoms, the cell geometry can not be expressed in term of atomic geometry.

Some properties of important crystals are presented below.

Table 2.1

Crystal Structure	*Side a and atomic radius r*	*Coordination Number*	*No of atoms/ unit cell*	*Packing Density*	*Examples*
Simple cubic	$a = 2R$	6	1	0.52	None
BCC	$a = 4R/\sqrt{3}$	8	2	0.68	Cr, Mo, W, α-Fe
FCC	$a = 4R/\sqrt{2}$	12	4	0.74	Ag, Au, Cu, Pt
HCP	$a = 2R$	12	2	0.74	Co, Mg, Ti, Zn
Diamond	$a = 8R/\sqrt{3}$	4	8	0.34	Diamond, Si, Ge
CsCl		8	1 cation 1 anion		Ionic solids CsCl, CsBr, CsI
Zinc blende		4	8	0.34	ZnS, GaAs, GaSb, InAs, InSb

Bravais Lattices There can be number of possibilities for choosing a unit cell for a given crystal structure. As a convention the geometry of a unit cell is represented by a parallelepiped with sides *a*, *b* and *c* and angles α, β, and γ as shown in figure. The sides *a*, *b*, *c* and angles α, β, and γ are referred to as lattice parameters. Based on different lattice parameters there are only seven possible distinct unit-cell geometries called crystal systems. These seven crystal system only categorize the unit cell based on geometry but do not mention where the atoms are placed within the cell. Depending upon the lattice arrays where the atoms are placed we can distinguish between similar crystal systems as in case of SC, FCC and BCC which all

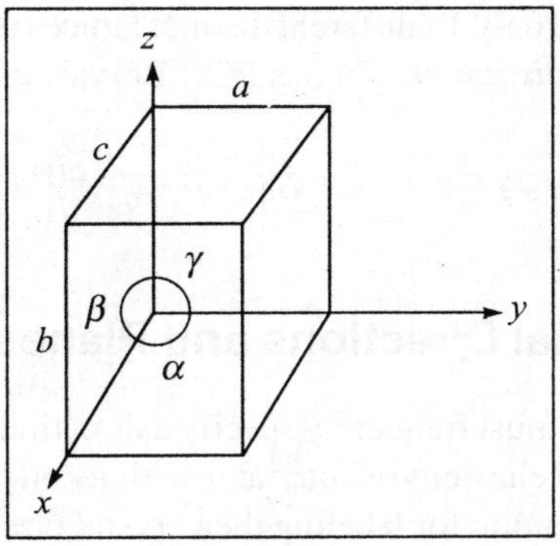

Fig. 2.4 Unit cell geometry

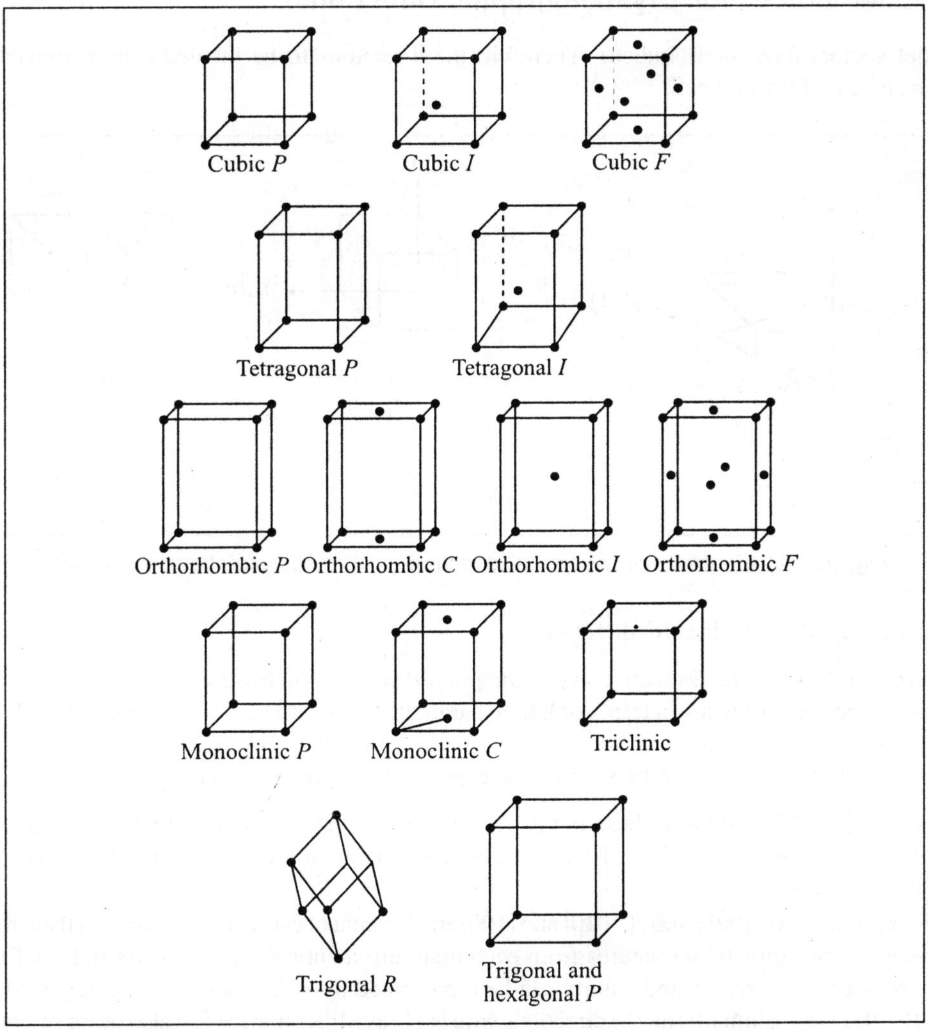

Fig. 2.5 Seven crystal systems and 14 Bravais lattices

have cubic crystal system. There are only 14 different distinct lattices that exist in three dimensions. These 14 distinct lattices are called ***Bravais lattice***. Cu has FCC Bravais lattice whereas Cr has BCC Bravais lattice.

For Cu and Fe, the unit cell geometry has $a = b = c$, $\alpha = \beta = \gamma = 90°$ and so cubic symmetry. For Zn $a = b \neq c$, and $\alpha = \beta = 90°$ and $\gamma = 120°$.

2.2 Miller Indices: Crystal Directions and Planes

In explaining crystal properties, we must frequently specify a direction and a plane of atoms in the crystal. Many properties such as modulus of elasticity, electrical resistivity, magnetic susceptibility are directional within a crystal. Here we use convention for labeling then crystal direction and plane.

2.2.1 Miller Indices for Crystallographic Direction

All parallel vectors have same indices. Therefore the directions to be labeled can be moved to pass through the origin of the unit cell.

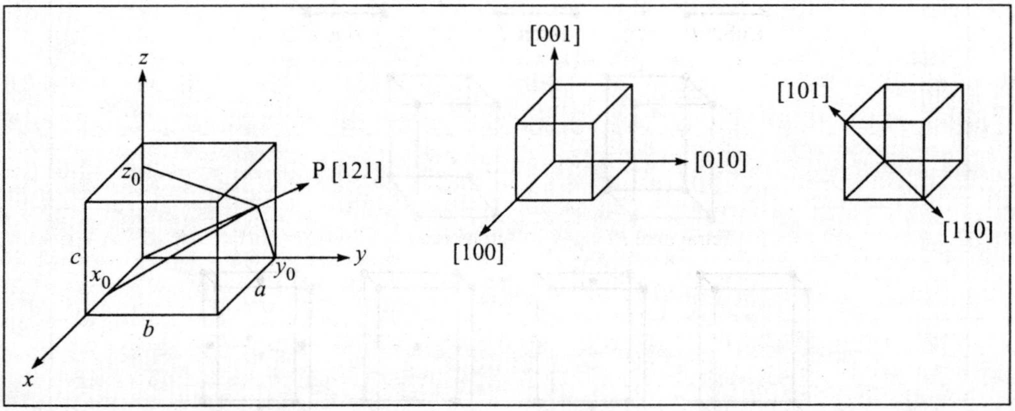

Fig. 2.6 (a) Identification of crystal direction (b) Various directions in crystals

Steps in finding Miller indices for directions:

- Find projections on the respective axes from point P (x_0, y_0, z_0 in Fig 2.6.b)
- Express these coordinates in terms of lattice parameters a, b and c x_0, y_0, z_0 are $a/2$, b, $c/2$
- P is at x_1, y_1, z_1 i.e. ½, 1, ½
- Multiply or divide these numbers till we have smallest integers including 0.

Here we will get 2,1,2. Putting these numbers without commas in square brackets, we will express required indices for direction as [212]. To indicate negative crystal direction we put a bar atop the number representing the index.

Directions [100], [010], [001], [001], [010] and [100] are different because of our assigning them respective axes. Otherwise their directional natures from each other are identical. All these six indices for crystal direction consisting two zeroes and a unity can be represented by <100> known as family of directions [100], [010], [001], etc. Similarly, all the directions with [111] as Miller indices can also be represented in the manner expressed above.

2.2.2 Miller Indices for Crystallographic Planes

The indices used to represent a crystallographic plane are called Miller indices for crystallographic planes. Here also the intercepts of the plane are expressed in terms of lattice parameters.

Steps in finding Miller indices for crystal planes:

- Find the intercepts of the plane on respective axes (x_0, y_0, z_0)
- Express these intercepts in terms of lattice parameters to obtain x_1, y_1, z_1
- Invert these numbers $1/x_1$, $1/y_1$, $1/z_1$
- Unlike in crystal directions, clear the fractions without reducing to lowest integers
- Put them in parentheses without commas.

For example, for the plane shown in figure 2.7.a, x_0, y_0, z_0 are (1/2)a, 1b and ∞c. Intercepts x_1, y_1, z_1 are ½, 1 and ∞. Their reciprocals are 2,1 and 0. There are no fractions, so (210) are the required Miller indices for crystallographic plane. In case there are negative numbers, bars atop these numbers are placed to distinguish them from positive indices. If a plane passes through origin we can shift it by a lattice parameter since planes shifted by a lattice parameter have same Miller indices. The planes represented by (100), (010), (001), (001), (010) and (100) are different due to our assigning them different axes. Otherwise their properties are exactly identical to each other. They are represented as family of planes <100>. Similar principle applies to planes with (111)s.

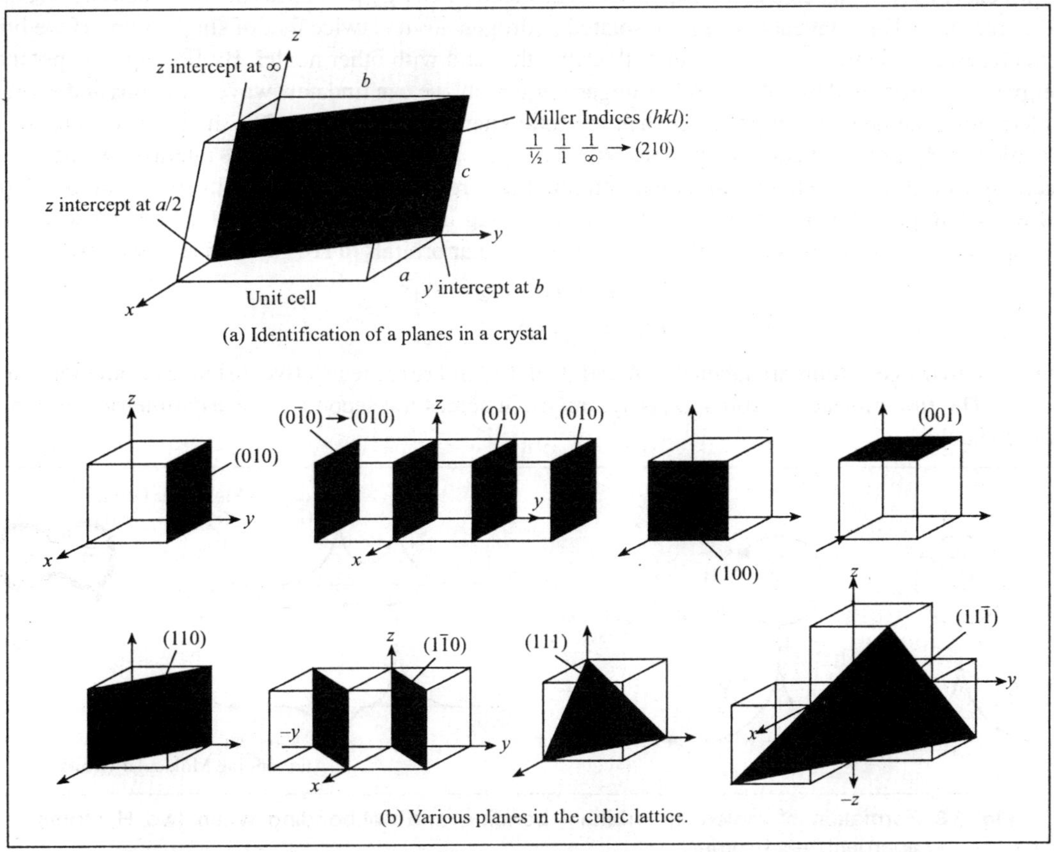

(a) Identification of a planes in a crystal

(b) Various planes in the cubic lattice.

Fig. 2.7 Labeling of crystal planes and typical examples in cubic lattice.

2.3 Band Theory of Solids

2.3.1 General Introduction

The free electron theory described in previous chapter can successfully explain the various properties of metal such as electrical and thermal conductivity, specific heat capacity, etc. However, the theory could not explain why certain solids behave as good conductors while certain others behave like insulators. It also could not explain why in certain solid, receptivity decreases with the increase in temperature.

The failure of the theory is due to the assumption of:

- constant or zero potential due to ion cores;
- the ion cores are free to move about in a crystal.

In fact, the potential due to ion cores is not constant rather it is periodic and the ion cores is at rest. These two reasonable considerations form the basis of the band theory of solids which results in the existence of band gap.

2.3.2 Molecular Orbital Bonding Theory

When two hydrogen atoms are very far apart so that they can not influence each others energy, each has its own energy levels, labeled as 1s, 2s, 2p, etc. The electron energy in each atom is –13.6eV with respect to the free state. The total energy of two isolated hydrogen atom is twice that of single atom. If we bring them closer, the electrons interact both with each other and with other nuclei. By finding new potential energy for electrons and by solving Schrödinger's equation, we can find new wave functions and energies of electrons. The new energy is lower than the total energy of electrons when at separate state, so the formation of H_2 is energetically favourable. When the ψ_{1s} orbital of both the atoms interact, two new wave functions with different energies and quantum numbers are formed. Two wave functions can overlap in two ways – in phase or out of phase. When they overlap in phase, ψ_{σ^*} will be formed and when they overlap out of phase ψ_{σ^*} will be formed. The two molecular orbitals in H-H system are given by

$$\psi_\sigma = \psi_{1s}(r_A) + \psi_{1s}(r_B)$$
$$\psi_{\sigma^*} = \Psi_{1s}(r_A) - \psi_{1s}(r_B) \qquad (2.2)$$

where two hydrogen atoms are labeled as A and B, and r_A and r_B are respective distances from their parent nucleus. The first molecular orbital ψ_σ is symmetric whereas the second ψ_{σ^*} is asymmetric and a node between nuclei.

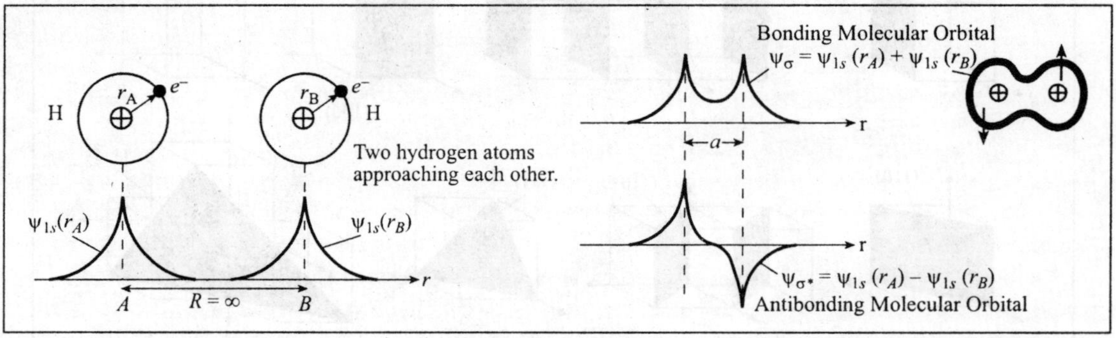

Fig. 2.8 Formation of molecular orbitals, bonding and antibonding when two H atoms approach each other

The wave functions of electrons in H_2 system must be determined by solving Schrodinger's equation, but we can guess that they must look like ψ_σ and $\psi_{\sigma*}$. The two energies E_σ and $E_{\sigma*}$ are widely different with E_σ below E_{1s} and $E_{\sigma*}$ above E_{1s}. The electrons are most likely to favour ψ_σ orbital as that would be the most stable outcome. The system would have lowest energy at $R = a$, where a is the separation between atoms, R-the atomic radius. The difference between energy of two isolated H atoms and minimum energy of H-H system is called bonding energy.

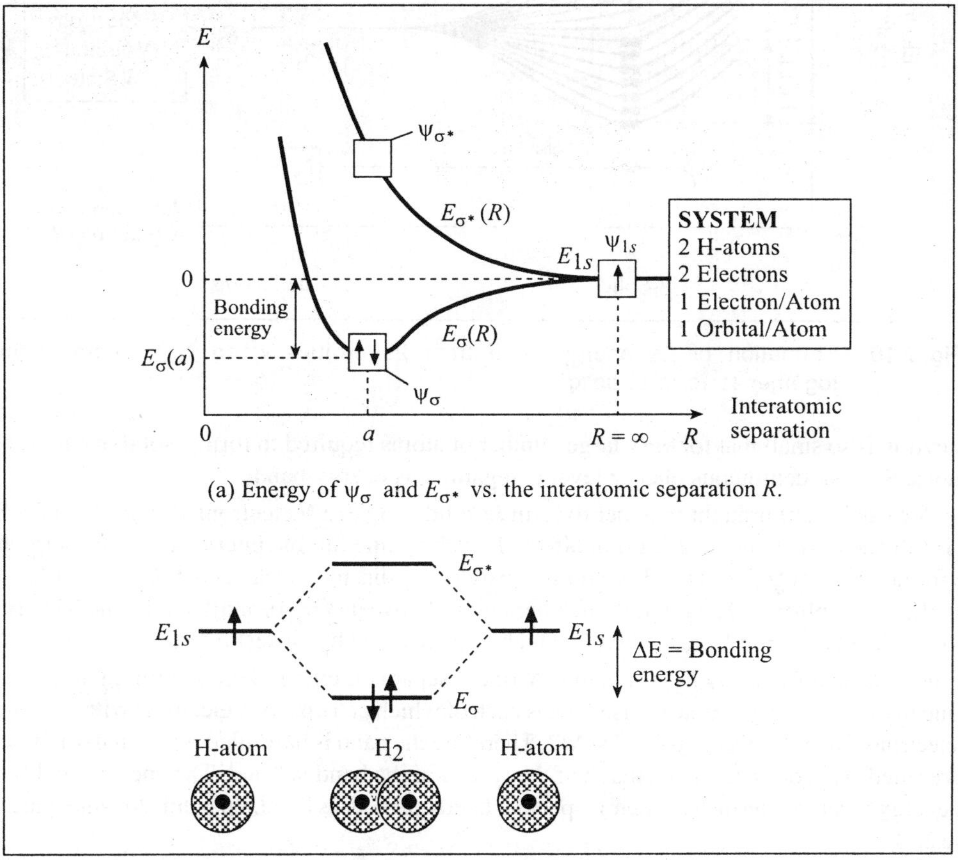

Fig. 2.9 Schematic diagram showing changes in electron energy as two isolated H atoms come together to form hydrogen molecule.

2.3.3 Band Theory of Solids

Alike in hydrogen atom, there are number of orbitals in atoms of different elements which at normal temperature give rise to a solid when they come together in large number. The identical orbitals of different atoms, with or without electrons within them interact (overlap) with each other and the out come will be as shown in figure 2.10. Some of the orbitals will have electrons occupying the state available where as some would have electrons occupying only the part of states available and others with no states occupied by electrons. When identical orbitals, for example ψ_{1s} or ψ_{2s} of different atoms overlap with respective orbitals, although the energy levels formed by overlapping of identical orbitals are finely separated from each other,

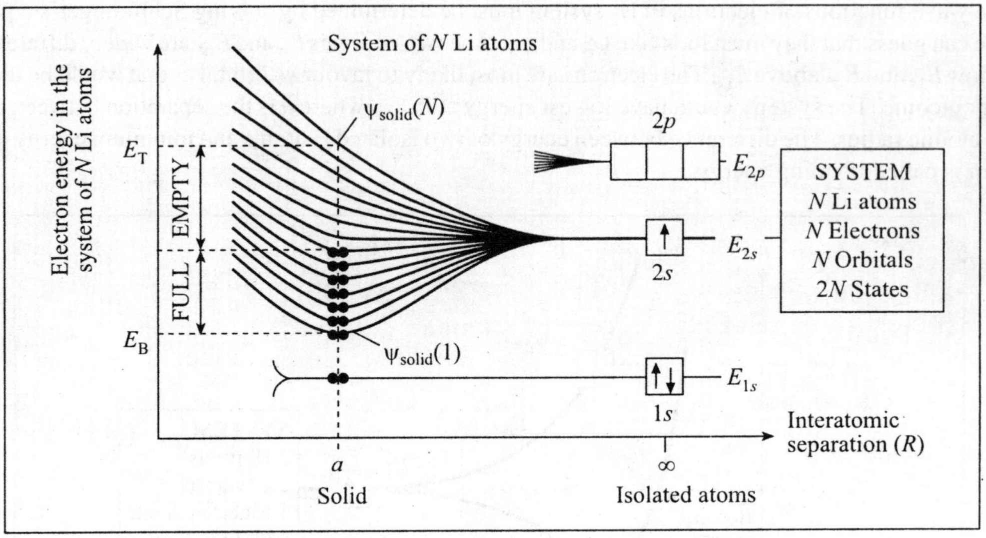

Fig. 2.10 Formation of 2s energy band from 2s orbitals when N Li atoms come together to form Li solid.

the separation is so small that for very large number of atoms required to form a solid, it can be assumed continuous. So, that continuous energy level is regarded as energy band.

When *N Li* atoms are brought together to form *Li* solid, we have *N* electrons in $N\psi_{2s}$ orbitals. Since 1s orbital is full and close to nucleus it is not affected much by interatomic interaction. The energy band E_{2s} will be formed from $N\psi_{2s}$ orbitals. The atoms separated by distance greater than $R = a$ will cause lesser amount of energy splitting. The interatomic interactions between $N\psi_{2s}$ orbitals spread the *N* energy levels from bottom level E_B to top E_T respectively which is determined by closest interatomic distance *a*.

The single 2*s* energy levels E_{2s} splits into *N* finely separated energy levels forming an energy band. Consequently there are *N* separate energy levels each of which can take two electrons with opposite spins. The *N* electrons fill all levels up to level at *N*/2. Therefore the band is half filled means half of the available states are filled. This does not mean that half the spread of the band is filled. The energy band formed by atomic energy levels are termed as their respective bands such as 2s band, 2p band, 3d band, etc as shown in figure 2.11.

As atoms approach each other's to form solids, at first, outermost energy levels are affected, and then only gradually, the inner or lower energy levels are affected. These close atoms' molecular orbitals interact resulting in new orbital of which only where the energy of the system is lower will be occupied by electrons. The splitting of energy level in very large finely separated energies that can be regarded continuous gives rise to what is called energy band. Not all the states available are filled, so with the energy bands also.

The splitting of single energy level of an isolated atom into an energy band is in the crystal lattice is responsible for various physical properties such as electrical, magnetic and optical.

In different solids, valence band is either completely or partially full. This is the highest occupied energy band with valence electrons. Valence band is the highest filled band. The next higher energy band is called conduction band, which can be completely of partially empty. Conduction band is the lowest unfilled band. The energy gap between valence band and conduction band is called forbidden gap where no states are available and so no electrons.

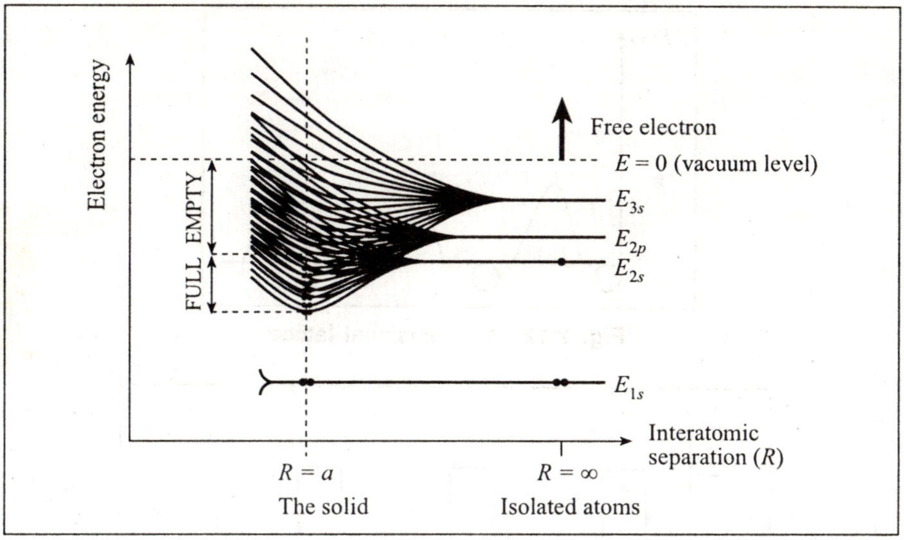

Fig. 2.11 Formation of different energy bands in Li solid due to large number of Li atoms brought close together.

The covalent forces of a crystal lattice are due to the valence band. If a valence electron absorbs enough energy, it can jump the energy gap and become a conduction electron after which it can freely move entirely through the material.

All the materials are classified on the basis of energy band gap. According to which there basically conductors, semiconductors and insulators. In conductors, there is virtually no energy gap between valence band and conduction band. They overlap with each other. The valence electrons can be readily assumed as conduction electron. In insulators, the energy band gap between valence band and conduction band is of the order of 10 eV. At absolute zero, all the states in valence band are completely filled and in conduction band completely empty. Even at room temperature there are very few electrons present in conduction band of an insulator. In semiconductors the energy band gap between valence band and conduction band is in average 1eV. Here also at absolute zero all the states in valence band are occupied and all the states in conduction band are empty. But at room temperature the number of electrons per unit volume in semiconductor is fairly higher than that of in insulators as less energy is required to excite valence electrons to conduction band in semiconductors compared with insulators.

2.3.4 Kronig-Penny Model

The free electron model implies that potential inside the solid is uniform in metals. A more accurate model allows for variation in potential energy due to the fixed positively charged lattice ions. In figure 2.12 a 1-dimensional lattice with ions separated by lattice parameter ***a***, is presented.

The highest potential is half way between the atom and goes to -∞ as position of the ion is approached. This potential distribution is quite complicated and for mathematical solution of Schrodinger's equation a simpler model known as "Kronig-Penny model" is used. The Kronig-Penny model still displays the following essential features where the function:

- has the same period as lattice
- potential is higher between ions and lower near the lattice ion

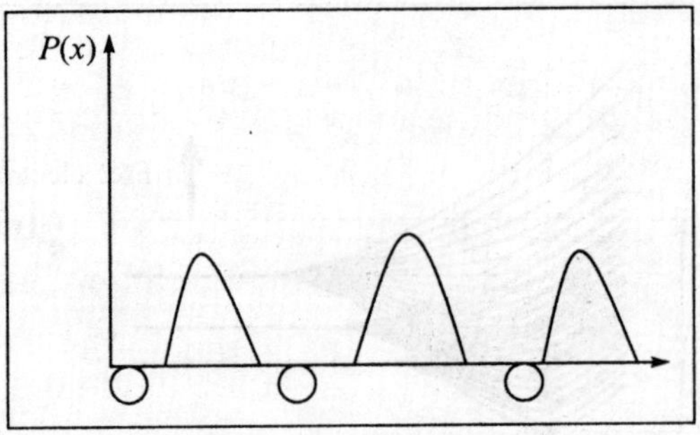

Fig. 2.12 1-dimensional lattice

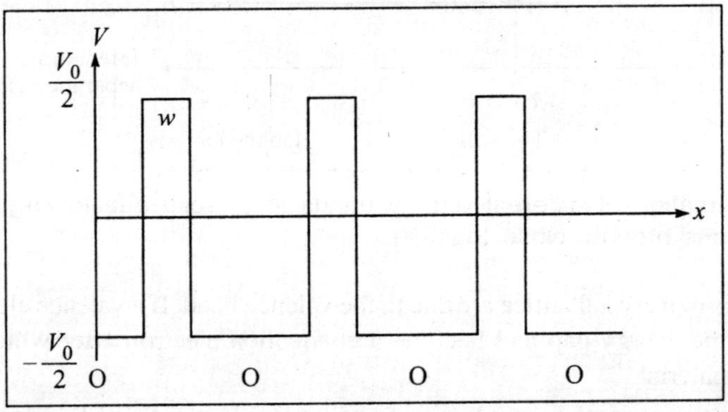

Fig. 2.13 Geometrical arrangement for Kronig-Penny model

The ions are located at $x = 0, a, 2a, 3a$, etc. The potential wells are separated from each other by potential barrier of height V_0 and width w. The Schrödinger's equation should be solved for potential distribution separately for $V = + V_0/2$ and $V = -V_0/2$ and compare the solutions at the boundary conditions.

The assumed wave function has the form

$$\psi(x) = U_k(x)e^{jkx} \tag{2.3}$$

where $U_k(x)$ is a periodic function with the same period as the lattice.

The solution for the above equation exists if k is related to energy E by the following equation and assuming that as $w \to 0$, $V \to \infty$, $wV = \text{const}$

$$\cos ka = \frac{P \sin \alpha a}{\alpha a} + \cos \alpha a \tag{2.4}$$

where
$$P = \frac{4\pi^2 ma}{h^2} V_0 w \tag{2.5}$$

$$\alpha = \frac{2\pi}{h}(2mE)^{1/2}$$

Here the relationship between E and k is not as easy as that of free electron. To find the exact E-k curve we must first of all plot the right hand side of equation (2.4) as a function of αa and then by plotting the left hand side and using the fact that this always must be between +1 to −1, a solution will be found for the values of E for which right side is within these limits.

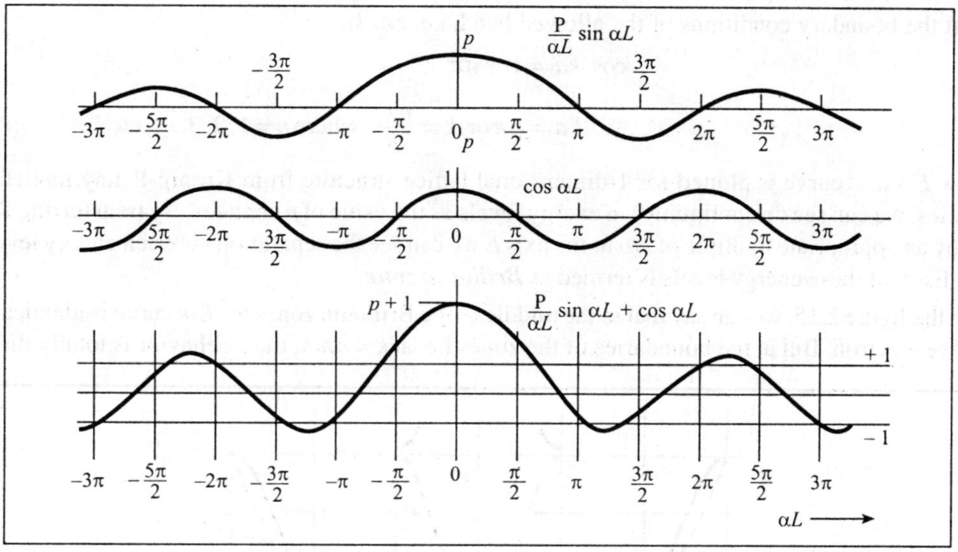

Fig. 2.14 Solution for Kronig-Penny model

The solution is defined only in the shaded region, i.e. only certain values are allowed. This implies that there are only certain allowed energy bands in crystal solids.

For the solution of Kronig-Penny model mentioned above, if the value of P is large i.e. $V_0 w$ is large, the function in given by right hand side of equation 2.4 crosses at a steep slope. This means that the allowed bands are narrower and forbidden bands are wider. Let us observe some specific conditions.

1. As $P \to \infty$, the band reduces to a single energy level, i.e. discrete energy spectrum for isolated atoms which is the actual case. So P is proportional to the potential barrier of the given system.

$$\frac{\sin \alpha a}{\alpha a} = 0$$

$$\sin \alpha a = 0$$

As $P \to \infty$, $\therefore$

$$\alpha a = n\pi$$

$$\frac{2\pi a}{h}\sqrt{2mE} = n\pi$$

$$E = \frac{n^2 h^2}{8ma^2}$$

which is exactly same as the equation for an electron in an infinite potential well of width a. So, all the electrons are independent of each other and confined to infinite potential barrier.

2. When $P \to 0$,

$$\cos \alpha a = \cos ka$$

$$\frac{2\pi a}{h}\sqrt{2mE} = ka$$

$$E = \frac{h^2 k^2}{8\pi^2 m} = \frac{h^2 k^2}{2m} \qquad (2.6)$$

Equation (2.6) is exactly same as that of equation for a free electron. So the variation in the value of P from zero to infinity accounts for the whole range of conditions in the crystalline solid.

3. At the boundary conditions of the allowed band, i.e. cos $ka = \pm 1$

$$\cos\, ka = \cos\, n\pi$$

$$ka = n\pi \text{ or, } k = \frac{n\pi}{a} \quad \text{where } n = 1, 2, 3, \ldots, \text{etc}$$

Below E vs. k^2 curve is plotted for 1-dimensional lattice structure from Kronig-Penny model. At the boundaries, we can see discontinuities in energy levels as the value of n changes. By transferring the zone curves by an appropriate multiple of π/a to the axis E we can see the separation between energy levels very clearly. Each of these energy levels is termed as ***Brillouin zone***.

From the figure 2.15, we can say that in the middle of the Brillouin zones the E-k curve is identical to that of the free electron. But at the boundaries of the zones i.e. at $k = n\pi/a$, their behavior is totally different.

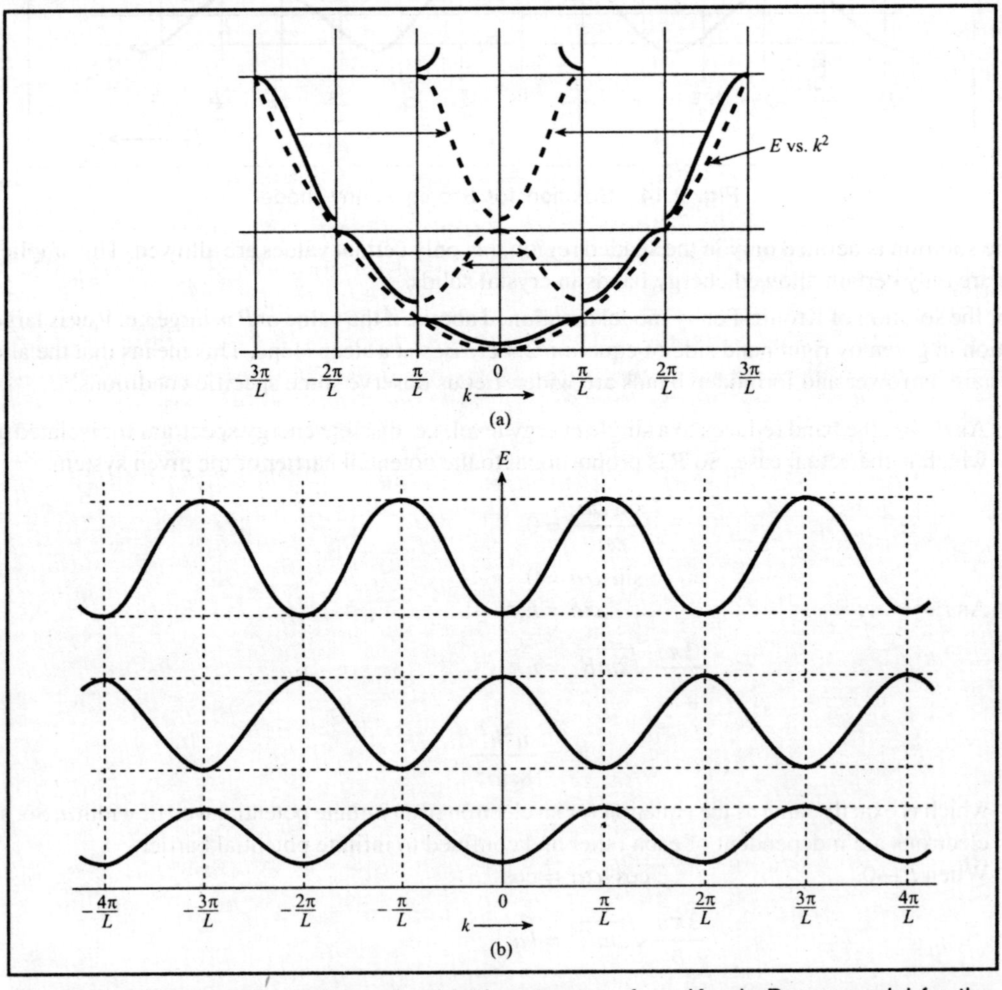

Fig. 2.15 E-k curve for 1-dimensional lattice structure from Kronig-Penny model for the normal and reduced zone cases. Also shown in the figure is the E-k curve for free electron. (b) plot of the extended zone curves for the same lattice structure.

Electrons that occupy Brillouin zone can move freely inside a crystal. All the information for the BZs are contained in the first BZ if we use the reduced zone plot. So, the BZs are the allowed energy regions in k-space, which corresponds to the allowed bands.

2.3.5 Effective Mass of an Electron

For a free particle the relationship between energy and momentum and correspondingly the wave number is

$$E = \frac{p^2}{2m} = \frac{h^2k^2}{2m} \tag{2.7}$$

For a particle in the crystal, energy in terms of k applies. Differentiating equation (2.7) with respect to k we will have

$$\frac{dE}{dk} = \frac{h^2 k}{m} \tag{2.8}$$

But the velocity, $v = \frac{p}{m} = \frac{hk}{m}$

And, comparing this with equation (2.8), we will get

$$\frac{1}{h}\frac{dE}{dk} = v \tag{2.9}$$

In equation (2.9) v is the velocity of particle in the crystal. Now acceleration is given by

$$a = \frac{dv}{dt} = \frac{d}{dt}\left(\frac{1}{h}\frac{dE}{dk}\right) = \frac{1}{h}\frac{d}{dt}\frac{dE}{dk}$$

$$\frac{d}{dt} = \frac{dk}{dt}\frac{d}{dk}$$

Also,
$$a = \frac{h}{m}\frac{dk}{dt}$$

So
$$\frac{h}{m}\frac{dk}{dt} = \frac{1}{h}\frac{dk}{dt}\frac{d^2E}{dk^2}$$

or,
$$m = \frac{h^2}{(d^2E/dk^2)} \tag{2.10}$$

Thus, mass is associated with curvature of energy in k-space. For a free particle this curvature $\frac{d^2E}{dk^2}$ is constant, so the E-k diagram is parabolic. But in the crystal solids this is not the case. The curvature $\frac{d^2E}{dk^2}$ is not always constant for bands representing E-k diagram for electrons within the solids.

For solids we can write

$$m^* = \frac{h^2}{(d^2E/dk^2)} \tag{2.11}$$

Where m^* represents the effective mass of the electron in crystal structure.

As the value ok k goes from $-\pi/a$ to 0, electron velocity changes from negative to zero. Similarly when k changes from zero to $+\pi/a$, velocity goes also changes from zero to positive values. The effective mass of the particle is positive for the curvature of $\frac{d^2E}{dk^2} \rangle 0$

i.e. the curvature is concave upward and the effective mass is negative when the curvature is concave downward for $\frac{d^2E}{dk^2} \langle 0$. For certain values of k the curvature is zero pointing toward the infinite mass. For $\frac{d^2E}{dk^2} \rangle 0$, mass is positive and electron is accelerated in the direction of the applied external force (due to external electric field) which matches with our perception of mass. In crystals, some times the electron is decelerated due to the external applied field. As external forces try to impel into a region dominated by a repulsive potential whose magnitude increases as the particle penetrates farther into the region, the result is what would be expected if the mass of the particle were negative. For certain cases, the electron cannot be accelerated whatever be the magnitude of the applied external force. The internal forces in the crystal cancel out the effect of external force. When electrons are accelerated indicating their positive masses, they can be accelerated beyond the zone boundary. When reaching the zone boundary, no further acceleration is possible whatever the applied force be. This can only be possible if the mass of the electron were infinite. This concept of velocity and mass of an electron in crystal is based upon perfect crystal structure, which is not the general case. There are always some crystal imperfections present in the structure.

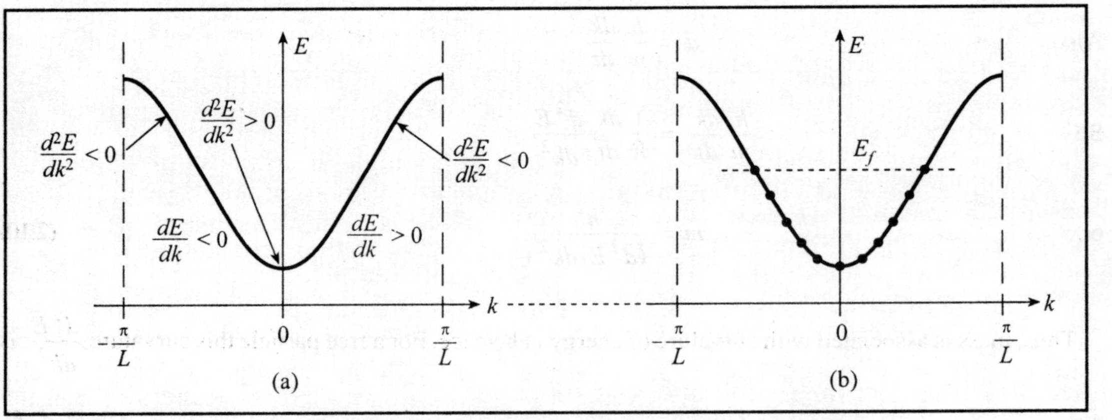

Fig. 2.16 Electron motion in the conduction band of a conductor for a simplified 1-dimensional structure (a) empty band; (b) at the equilibrium.

2.4 Thermal Velocity of Electron at Equilibrium

Applying kinetic molecular theory of gases for gas molecules contained in a container we can write the final theoretical derivation in the form

$$PV = \frac{Nmv^2}{3} \tag{2.12}$$

where P = Pressure of gas inside the container
V = Volume of the container
N = Number of gas molecules in container
m = Mass of the gas molecule
v^2 = mean square velocity of molecules

The experimental observation is

$$PV = \frac{N}{N_A} RT = nRT \tag{2.13}$$

where N_A = is Avogardo's number
R = Universal gas constant
T = Temperature
$n = N/N_A$ = Number of moles of gas

For N molecule gas, every molecule will have its random velocity. The change in momentum of the molecule following collision with the wall of the container is $\Delta p_x = 2mv_x$ where m is the molecular mass. Time interval between collision will be $\Delta t = 2a/v_x$ where a is the side of container. Force exerted by this molecule on the face A of the container is $F_x = \frac{\Delta p}{\Delta t} = \frac{mv_x^2}{a}$.

And the total pressure exerted by N molecules on face A due to total force is

$$P = \frac{\text{Total Force}}{\text{Area}} = \frac{mv^2{}_{x1} + mv^2{}_{x2} + \ldots + mv^2{}_{xN}}{a^3} = \frac{m}{a^3}(v^2{}_{1x} + v^2{}_{x2} + \ldots + v^2{}_{xN})$$

$$P = \frac{mN}{V} v_x^2$$

where $Nv_x^2 = v^2{}_{x1} + v^2{}_{x2} + \ldots + v^2{}_{xN}$

Since the molecules are in random motion and collide randomly with each other, mean square velocity in x direction will be equal to that in y and z directions.

So, $$v_x^2 = v_y^2 = v_z^2$$

The velocity of any molecule is given by

$$v^2 = v_x^2 + v_y^2 + v_z^2 = 3v_x^2$$

v^2 = mean square velocity of the molecule.

We will find that average kinetic energy per degree of freedom is (can be derived) $kT/2$. So for three degrees of freedom this is $3kT/2$. Average kinetic energy per mole is $mv^2/2$.

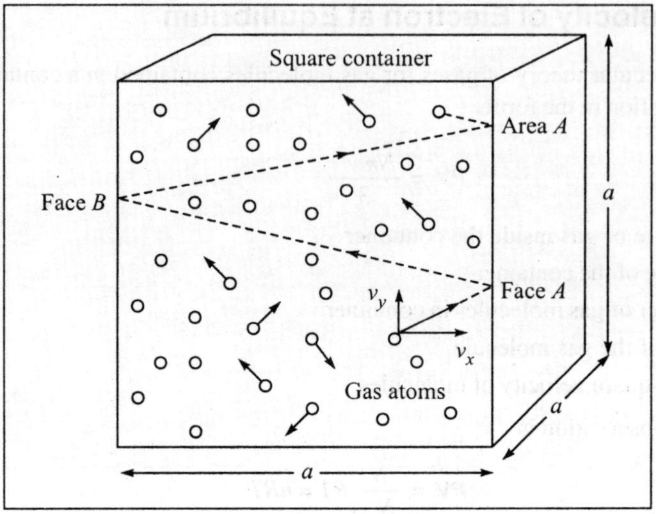

Fig. 2.17 The gas molecules in the container

So, $$KE = \frac{mv^2}{2} = \frac{3}{2}kT$$

$$v_{th} = (3kT/m)^{1/2} \tag{2.14}$$

where v_{th} is the mean thermal velocity of electrons.

The equilibrium is maintained through the interaction with the lattice. Lattice vibrations called ***phonons*** are the result. Thermal energy possessed by lattice atoms causes vibrations about the equilibrium positions of regular periodic array. These distortions are called phonons. The energy exchange between electron and lattice crystal is called scattering.

2.5 Electron Mobility, Conductivity and Resistivity

Although the electrons are in thermal random motion, no net flow of electron in a particular direction persists in the absence of external electric field. So to maintain the flow of electron we need to apply an external electric field. V is the applied potential across the length L of the conductor. Then the electric field will be given by

$$E = \frac{V}{L} \tag{2.15}$$

This electric field is present at every point of the solid. The application of this field will result in acceleration of the electron given by

$$F = m^* a$$

$$F = -eE$$

So, $$a = \frac{(-e)}{m^*}E \tag{2.16}$$

Thus in addition to random velocity, electron acquires a velocity in the direction of applied field. This directed velocity is lost after each collision with the lattice atoms because of the fact that electron is very

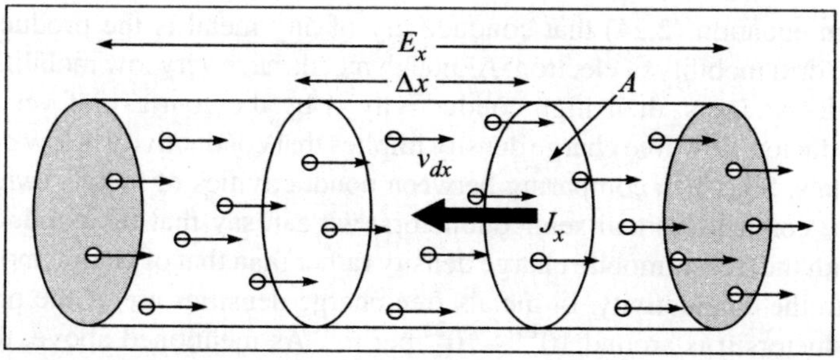

Fig. 2.18 Drift of electrons in a conductor in the presence of an applied electric field

light compared to the atom. Therefore, only the part of the velocity that is picked up between collisions actually counts. The average time between two collisions is termed as ***collision time*** or ***mean free time*** or ***relaxation time***.

It is denoted by Greek alphabet tau (τ). If τ is the average time between collision, then the final velocity of electron will be $a\tau$. So the average velocity is

$$v_{av} = a\tau/2 \tag{2.17}$$

A more careful technique for averaging the velocities results in the replacement of the factor 2 from the denominator of equation (2.17) with unity.

So
$$v_{av} = a\tau \tag{2.18}$$

V_{av} is called drift velocity of electron.

From equations (2.16) and (2.18) and neglecting the negative sign in (2.16), we have

$$v_D = (e\tau/m)E \tag{2.19}$$

$$v_D = \mu_D E \tag{2.20}$$

where v_D = drift velocity of electron in crystalline solid

μ_D = drift mobility of electron

Drift mobility is related to mean free time and mass of electron as

$$\mu_D = \frac{e}{m}\tau \text{ cm}^2\text{V}^{-1}\text{s}^{-1} \tag{2.21}$$

Assuming that all the electrons drift with v_D, the number of electrons crossing a plane of unit area per second can be obtained by multiplying the drift velocity by the free electron density N_e.

$$\text{Electron Flow} = N_e v_D \text{ s}^{-1}\text{m}^{-2}$$

$$\text{Current Density} = \text{Charg ex Flow}$$

$$J = N_e v_D e$$

But
$$v_D = \mu_D E = \frac{e}{m}\tau E$$

So,
$$J = eN_e \frac{e}{m}\tau E = \frac{e^2 N_e \tau}{m} E \tag{2.22}$$

$$J = \sigma E \tag{2.23}$$

where σ = Electrical conductivity

$$\sigma = \frac{e\tau}{m} eN_e = \mu_D e N_e \quad \text{Sm.m}^{-1}\text{(read: Siemens per meter)} \tag{2.24}$$

So, it is clear from equation (2.24) that conductivity of any metal is the product of charge density (electron density) and drift mobility of electron. Although metals have very low mobilities approximately 1/100th of that of semiconductors, their high conductivity is by the courtesy of very high free electron densities. In semiconductors, low free charge density implies that conductivity is low despite high mobility compared to the metals. So, when comparing between conductivities of metals (which are called good conductors or simply conductors) and semiconductors, we can say that the conductivity difference is mainly associated with the free or mobile charge density rather than that of charge mobility although both are directly related to the conductivity. In metals free charge densities are of the order of 10^{28} per m^3, whereas in semiconductors it is around $10^{16} - 10^{22}$ per m^3. As mentioned above, the variation in drift mobility is not as high as that of free charge density so as to compensate the latter's effect.

The reciprocal of the conductivity is termed as Resistivity of the material.

$$\rho = \frac{1}{\sigma}, \Omega\text{.m (read: ohm meter)} \tag{2.25}$$

Resistance of a conductor with cross-sectional area A and length L is calculated as

$$R = \rho\frac{1}{A} \tag{2.26}$$

When calculating resistance, there should be no confusion between the usual meaning of length and the length of current path. In usual meaning, the longest dimension is referred as length whereas for electrical resistance calculation, the length of the path along which current flows is taken as length. In the similar manner, the cross-sectional area of the current path is calculated.

2.6 Physical Properties of Some Common Metals

Table 2.2 Resistivity, thermal coefficient of Resistivity at 273° K for various metals. n is the index in $\rho \propto T^n$.

Metal	ρ_o, $10^{-9}\,\Omega$ m	α_o, K^{-1}	n	*Comments*
Aluminum	25	1/233		
Copper	15.7	1/232	1.15	
Gold	22.8	1/251		
Antimony	38	1/196		
Platinum	98	1/255	0.94	
Silver	14.6	1/244	1.11	
Tantalum (Ta)	117	1/294	0.93	
Tungsten	50	1/202	1.2	
Tin	110	1/217	1.11	
Iron	84	1/152	1.8	Magnetic 273K<T<1043K
Nickel	59	1/125	1.72	Magnetic 273<T<627K

2.7 Solved Numerical Examples

1. Calculate the root mean square velocity of nitrogen molecule in atmospheric air at 27°C. Also calculate the *rms* velocity in one direction.

Solution:

We have
$$\frac{mv^2_{rms}}{2} = \frac{3kT}{2}$$

$$v_{rms} = \left[\frac{3kT}{m}\right]^{1/2}$$

where m is the mass of nitrogen molecule N_2, k – Boltzman's universal constant.
$k = 1.38 \times 10^{-23}$ J/K and the atomic mass of N is M_{at} = 14g/mol

Avogardo's Number, $N_A = 6.022 \times 10^{23}$ mol^{-1}

Mass of the molecule is related to atomic mass and N_A as

$$m = \frac{2M_{at}}{N_A} = \frac{2 \times 14 \times 10^{-3} \text{ kg/mol}}{6.022 \times 10^{23}/\text{mol}}$$

And the required *rms* velocity is

$$v_{rms} = \left[\frac{3 \times (1.38 \times 10^{-23} \text{ J/K})(300\text{K})}{(28 \times 10^{-26}/6.022\text{kg})}\right]^{1/2}$$

$$v_{rms} = 517 \text{ (J/kg)}^{1/2} = 517[\text{kg (m/s)}^2/\text{kg}]^{1/2}$$
$$= 517 \text{ ms}^{-1}$$

Now, velocity in one direction (for example x-direction) can be found as

$$v^2_{rms} = v^2_{rmsx} + v^2_{rmsy} + v^2_{rmsz}$$

The average velocities in all three directions are supposed to be equal to each other in magnitude. So,

$$v_{rmsx} = \sqrt{\frac{v^2_{rms}}{3}} = \frac{v_{rms}}{\sqrt{3}} = 298\text{m/s}$$

2. Calculate the density of copper with $a = .362$nm, $M_{at} = 63.5$g/mol. $N_A = 6.022 \times 10^{23}$/mol.

Solution:

Copper has FCC structure and 4 atoms per unit cell.

$$\text{Density } \rho = \frac{m_{unitcell}}{V_{unitcell}} = \frac{4 \times (63.5 \times 10^{-3} \text{ kg/mol})/(6.022 \times 10^{23}/\text{mol})}{(0.362 \times 10^{-9} \text{ m})_3}$$

$$= 8890\text{kg/m}^3$$

3. Calculate the drift mobility of electrons in copper at room temperature. Conductivity, mass and density of copper are 5.9×10^5 Sm/cm, 63.5 g/mol and 8.93 g/cm^3 respectively.

Solution:

Drift mobility is given by $\mu_d = \sigma/en$ where n is the electron concentration in the material. From Avogardo's number, we know that one mole of copper has 6.022 × 10^{23} atoms and each Cu atom contributes one electron to the conduction band. The mass of one mole of copper is 63.5g. So, from this concentration of electrons in copper can be found as

$$n = \frac{\rho N_A}{M_{at}} = \frac{(8.93\ \text{g/cm}^3)\ (6.022 \times 10^{23}/\text{mol})}{(63.5\text{g/mol})}$$

$$= 8.5 \times 10^{22}/\text{cm}^3$$

Now, the drift mobility is $\frac{(8.93\ \text{g/cm}^3)\ (6.022 \times 10^{23}/\text{mol})}{(63.5\text{g/mol})}$

$$\mu_d = \frac{\sigma}{en} = \frac{(5.9 \times 10^5\ \text{Sm/cm})}{(1.6 \times 10^{-19}\ \text{C})\ (5.5 \times 10^{22}/\text{cm}^3)}$$

$$\mu_d = 43.4\text{cm}^2\text{V}^{-1}\text{s}^{-1}$$

which is more than experimental value of $32\text{cm}^2\text{V}^{-1}\text{s}^{-1}$ (Hall effect) and due to the fact that more than one electron is contributed to the conduction band by each atom.

4. What applied electric field can impose a drift velocity equal to 10% of mean speed ~$1.2 \times 10^6\ \text{ms}^{-1}$ of the conduction electrons in copper? Use the experimental value of drift mobility.

Solution:

Applied electric field is related to drift velocity and drift mobility as

$$E = \frac{v_d}{\mu_d} = \frac{1.2 \times 10^6 \times 0.1\text{ms}^{-1}}{32\text{cm}^2\text{V}^{-1}\text{s}^{-1}} = 3.8 \times 10^5\ \text{Vcm}^{-1}$$

$$= 3.8 \times 10^7\ \text{Vm}^{-1} = 3.8\ \text{MVm}^{-1}$$

So, for even the highest working voltages, the drift velocity is much smaller than the mean speed of the electrons. Consequently, when an electric field is applied to a conductor, mean speed is unaffected for all practical purpose.

5. The width of energy band is typically ~10eV.

Calculate:

(i) The density of states at the center of the band;
(ii) ii. The number of states per unit volume within a small energy range *kT* about the center;
(iii) iii. The density of states *kT* above the center;
(iv) iv. The number of states per unit volume within a small energy range of *kT* to 2*kT* from the bottom of the band.

Solution:

The density of states or the number of states per unit energy range per unit volume considering Pauli's exclusion principle is

$$Z(E) = 8\pi\sqrt{2}\,\frac{m_e^{3/2}}{h^3}\,E^{1/2}$$

$$= 8\pi\sqrt{2}\,\frac{(9.1 \times 10^{-31}\ \text{kg})^{3/2}}{(6.624 \times 10^{-34}\ \text{Js})^3}\,(5 \times 1.6 \times 10^{-19}\ \text{J})^{1/2} = 9.5 \times 10^{46}\ \text{m}^{-3}\ \text{J}^{-1}$$

In per cm^3 per eV,

$$Z(E) = (9.5\times 10^{46}\,\text{m}^{-3}\,\text{J}^{-1})\,(10^{-6}\,\text{m}^3\,\text{cm}^{-3})\,(1.6\times 10^{-19}\,\text{JeV}^{-1})$$
$$= 1.52\times 10^{22}\,\text{cm}^{-3}\,\text{eV}^{-1}$$

To find the number of states per unit volume within kT about the center of the band we will multiply the density of states at the center by $E = kT = 0.0259$eV at 300K.

$$N(E) = Z(E)\,E = 1.52\times 10^{22}\text{cm}^{-3}\text{eV}^{-1}\times 0.0259\text{eV} = 3.9\times 10^{20}\,\text{cm}^{-3}$$

$$Z(E)kT = 8\pi\sqrt{2}\,\frac{(9.1\times 10^{-31}\text{ kg})^{3/2}}{(6.624\times 10^{-34}\text{ Js})^3}(0.0259\times 1.6\times 10^{-19}\text{ J})^{1/2} = 6.84\times 10^{45}\,\text{m}^{-3}\,\text{J}^{-1}$$
$$= 1.1\times 10^{21}\,\text{cm}^{-3}\,\text{eV}^{-1}$$

Within the kT the number of states is

$$N(E) = 1.1\times 10^{21}\,\text{cm}^{-3}\,\text{eV}^{-1}\times 0.0259\text{eV} = 2.8\times 10^{19}\,\text{cm}^{-3}.$$

So, even at the bottom of the band, the concentration is fairly high.

6. For silver with $E_{F0} = 5.5$eV and $\Phi = 4.5$ eV, calculate the total number of states per unit volume and compare this with atomic concentration of silver. Density and atomic mass of silver are 10.5g/cm^3 and 107.9 g/mol respectively.

Solution:

For silver with Fermi energy at absolute zero = 5.5eV and work function = 4.5eV, the width of the energy band $E = 5.5 + 4.5 = 10$ eV. The density of states function reaches maximum at the center of the band. From integrating from bottom of the band to center of the band, we will find number of states per unit volume up to the center of the band which is half the total number of states.

$$\frac{1}{2}N(E) = \int_0^5 Z(E)dE = \int_0^5 8\pi\sqrt{2}\,\frac{m_e^{3/2}}{h^3}E^{1/2}dE$$

$$\frac{1}{2}N(E) = \frac{16\sqrt{2}\pi m_e^{3/2}}{3h^3}(5-0)^{3/2} = 5.08\times 10^{28}\,\text{m}^{-3}$$

So,

$$N(E) = 10.16\times 10^{28}\text{ states/m}^3$$

Now,

$$\text{Density } \rho = \frac{nM_{at}}{N_A}$$

$$n = \frac{\rho N_a}{M_{at}} = \frac{(10.5\times 10^{-3}\text{ kg}/10^{-6}\text{ m}^3)\,(6.022\times 10^{23}/\text{mol})}{107.9\times 10^{-3}\text{ kg/mol}}$$

$$n = 5.85\times 10^{28}\text{ atoms/m}^3$$

So, even with a crude approximation, as expected, the number of states is double the atomic concentration.

7. Calculate the Fermi energy at 0K for copper. Given its density is 8.96 g/cm^3 and atomic mass 63.5g/mol. $N_a = 6.022\times 10^{23}$/mol.

Solution:

Assuming each atom contributes one electron to the conduction band, the electron concentration of copper is

$$n = \frac{\rho N_A}{M_{at}} = \frac{(8.96\text{g}/\text{cm}^{-3})\,(6.022\times10^{23}/\text{mol})}{(63.5\text{g}/\text{mol})} = 8.5\times10^{22}\ \text{cm}^{-3}$$

The Fermi energy at 0K is given by

$$E_{F0} = (h^2/8m_e)\,(3n/2\pi)^{2/3}$$

$$E_{F0} = \frac{(6.624\times10^{-34}\ \text{Js})^2}{8\times9.1\times10^{-31}\ \text{kg}}\left(\frac{3\times8.5\times10^{28}\ \text{m}^{-3}}{2\pi}\right)^{2/3} = 1.1\times10^{-18}\,\text{J} = 7\text{eV}$$

8. Drift mobility of conduction electron is 43cm^{2}V^{-1}s^{-1} and mean speed is 1.2×10^6 ms^{-1}. Calculate the mean free path of electrons between collisions.

Solution:

The mean free path between collisions is given by

$$L_e = v_e\tau$$

where v_e is the electron velocity,

τ - is the mean free time between collisions.

But, $\mu_d = \dfrac{e\tau}{m_e}$ $\tau = \dfrac{\mu_d m_e}{e}$

So, $L_e = v_e \dfrac{\mu_d m_e}{e} = 1.2\times10^6\ \text{ms}^{-1}\dfrac{(43\text{cm}^2\text{V}^{-1}\text{s}^{-1})\,(9.110^{-31}\text{kg})}{1.6\times10^{-19}\,\text{C}}$

$L_e = 2.99\times10^{-8}\text{m} = 29.9\text{nm}$

Exercise

1. Calculate the *rms* velocity of oxygen molecules at 127°C at a pressure of 2×10^5Nm^{-2}. [558ms^{-1}]
2. Calculate the number of electrons per unit volume of copper. Molar mass of copper = 63.5 and Avogardo's number = 6.022×10^{23}/mol. [8.45×10^{28}m^{-3}]
3. Calculate the Fermi kinetic energy at absolute zero temperature for valence electrons in Na and Li. Why does the Fermi energy not depend on the size of the piece of a given metal? Assume $m^* = m$. [3.11eV and 4.72eV]
4. Calculate the number of atoms per unit cell of a metal having lattice parameters of 2.9Å and density of 7.87 g/cc. Atomic weight of the metal is 55.85 and $N_A = 6.023\times10^{23}$/mol. [2]
5. If the electrical conductivity of potassium is 1.39×10^5 Sm/cm, calculate the drift mobility of electrons at room temperature. Molar mass and density of potassium are 39.95 and 0.91 g/cc. [61.96 cm^{2}V^{-1}s^{-1}]
6. Show that for cubic lattice, the lattice constant a is given by $a = \left[\dfrac{NM}{N_A\rho}\right]^{1/3}$

   where N is the number of atoms per unit cell, M is atomic weight, N_A is Avogardo's number, and ρ is density of the crystal material.

7. Show that for any cubic lattice, the separation of the planes corresponding to the Miller indices $(hk1)$ is given by $d_{hk1} = \dfrac{a}{\sqrt{h^2 + k^2 + l^2}}$

8. Platinum having molar mass 195.1crystallizes in the FCC structure and has a density of $21.4 \times 10^3 \text{Kg.m}^{-3}$. Calculate the length of the edge of the unit cell. [0.392nm]

9. Using the Kronig – Penny model, show that the energy of the lowest energy band is $E = \dfrac{h^2 P}{ma^2}$ where p<< 1.

10. Zinc has the crystal structure HCP. At room temperature, its unit cell parameters are $c = 4.947$ Å and $a = 2.665$ Å. The bulk density of zinc is 7134 Kg.m^{-3}. Determine the atomic weight of zinc. [65.38]

11. A copper wire of cross-sectional area of 5mm^2 carries a steady current of 50A. Calculate (a) density of free electrons and (b) average drift velocity of electrons. (Hints: Assume one free electron per atom) [(a) $8.37 \times 10^{28} \text{m}^{-3}$ (b) $7.47 \times 10^{-4} \text{m.s}^{-1}$]

12. Prove that the conductivity of a metal is inversely proportional to its absolute temperature.

13. The valence band of the simple cubic metal has the form $\boldsymbol{E = AK^2 + B}$, where $A = 10^{-3} \text{J.m}^2$ and B = – 12eV. The band is a p-band states up to $K_a = (2\pi^2)^{1/3}$

Conduction in Liquids and Gases

3 Chapter

3.1 Introduction

We are mostly familiar with many liquids and gases being used as dielectric materials in different applications in electrical engineering to ensure the required insulation level for proper functioning of different electrical appliances, devices and machines. But, in addition to the abovementioned, we also encounter many liquids and gases used for conduction purposes as well. In certain special circumstances, we are forced to use liquids and gases to ensure the conduction process takes place smoothly as demanded by the circumstance. There maybe such special cases, where the same material is meant for dual purposes, firstly act as insulation and then during special conditions, act as conducting medium.

3.2 Ionic Conduction in Liquids

The practical aspect of ionic conduction in electrolytes can be visualized from the examples purposely done electrolysis electroplating electrolysis of water and naturally occurring phenomena.

In case of purposely done electrolysis or electroplating, what we do is to take aqueous media containing the metal to be plated in which two electrodes called cathode and anode are placed. When an external potential is applied, cation and anion are formed, and cation is attracted towards anode while the anion goes towards cathode.

In case of naturally occurring ionic conduction due to surrounding environmental conditions, aqueous media is formed. And, due to electron affinity of material, ionization takes place.

One of the examples of ionic conduction in liquids is electroplating in which following reaction takes place:

$$CuSO_4 \rightarrow Cu^{++} + SO_4^{--} \tag{3.1}$$

$Cu^{++} \rightarrow$ goes to anode looses positive charge and sticks on to the anode;
$SO_4^{--} \rightarrow$ goes to cathode which is made up of copper, so knocks copper to form $CuSO_4$ in solution making the concentration same as before.

Similar examples of electroplating of other metals and the electrolysis of water can also be recalled in which ionic conduction takes place.

In natural form of ionic conduction is exhibited by rusting of iron. Iron in free environment due to the humidity of the air, forms an aqueous environment is followed by anodic reaction in the following form:

$$Fe^0 \rightarrow Fe^{2+} + 2e^- \tag{3.2}$$

$$\text{Cathode reaction } Fe^{2+} + 2e^- \rightarrow Fe^0 \tag{3.3}$$

All the above mentioned processes take place in the following manner. In ionic conduction the individual atoms transfer electrons between each other to form positively charged cation and negatively charged anions. The binding forces between the ions are electrostatic in nature and are thus very strong. The room temperature conductivity of such liquid materials are very low in comparison to conductivity of metals because only limited number of electrons from valance band can reach the conduction band and result in ionization.

The main conduction is caused due to movement of such positively or negatively ion in particular direction on application of electric field. The ionic conductivity is given by

$$\sigma_{ionic} = eN_{ion}\mu_{ion} \tag{3.4}$$

where N_{ion} is the number of ions, which are ionized, and moving under the influence of applied electric field. And anion is the mobility of these ions.

In order to move, these ions must possess sufficient kinetic energy and electric field. During the movement of these ions, there is collision and exchange of energy. Hence ionization may be increasing successively and directional in nature due to applied electric field.

From Einstein diffusion relation, we have

$$\mu_{ion} = \frac{eD}{kT} \tag{3.5}$$

where D is the diffusion coefficient for ions; k is Boltzmann's constant and T is the temperature of the liquid.

The diffusion coefficient varies with temperature as

$$D = D_0 \exp(-Q/kT) \tag{3.6}$$

where Q is the activation energy for the purpose under consideration, and D_0 is the pre-exponential factor which depends on the vibrational frequency of the atoms and other material properties.

Ionic conductivity can be expressed as

$$\begin{aligned}\sigma_{ion} &= eN_{ion}\mu_{ion}\\ &= eN_{ion}\frac{eD}{kT}\\ &= eN_{ion}\frac{e}{kT}D_0\exp(-Q/kT)\\ &= \frac{e^2N_{ion}}{kT}D_0\exp\left(-\frac{Q}{kT}\right)\end{aligned}$$

i.e.

$$\sigma_{ion} = \sigma_0 \exp\left[-\frac{Q}{kT}\right] \tag{3.7}$$

where

$$\sigma_0 = \frac{e^2N_{ion}}{kT}D_0 \tag{3.8}$$

Taking natural logarithm on both the sides, we get

$$\ln \sigma_{ion} = \ln \sigma_0 - \frac{Q}{kT} = \ln \sigma_0 - \left(\frac{Q}{k}\right)\frac{1}{T} \tag{3.9}$$

The above relation gives the dependence of ionic conductivity on the inverse of temperature.

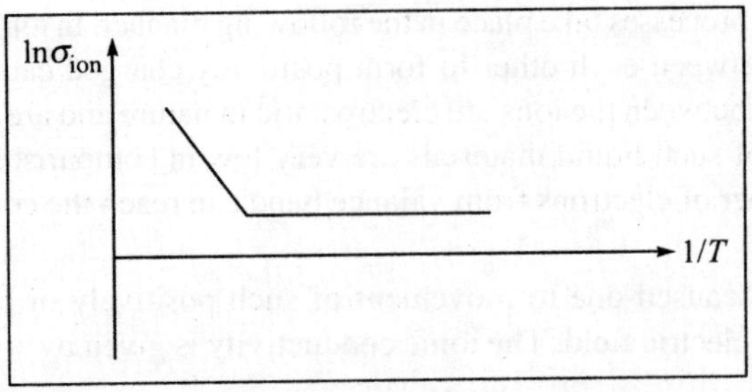

Fig. 3.1 Dependence of ionic conductivity on inverse of temperature.

In addition to the above-mentioned ionic conduction in electrolytes, conduction also takes place in liquid dielectric materials. When low electric fields of less than 1kV/cm are applied the conductivity of the liquid ranges from 10^{-18} to 10^{-20}Sm/cm due to the presence of impurities in the dielectric after the purification process. However, when the fields are very high ~100kV/cm, there is rapid rise in current with violent fluctuations. This conduction current shown in figure 3.2 is nearer to breakdown of dielectric.

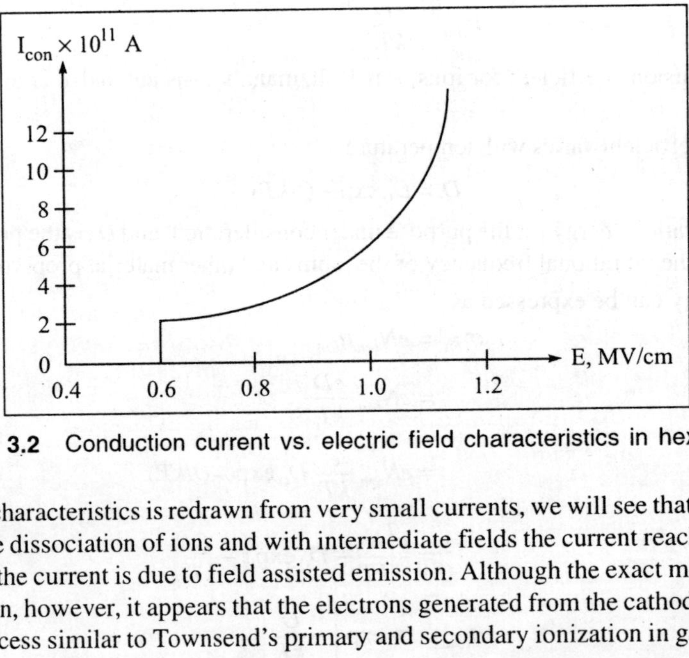

Fig. 3.2 Conduction current vs. electric field characteristics in hexane

However if this characteristics is redrawn from very small currents, we will see that at initial stages, the current is due to the dissociation of ions and with intermediate fields the current reaches saturation value and for high fields the current is due to field assisted emission. Although the exact mechanism of current growth is not known, however, it appears that the electrons generated from the cathode field emission get multiplied by a process similar to Townsend's primary and secondary ionization in gases.

Commercial liquid insulating materials such as hexane, benzene, transformer oil, silicone, liquid nitrogen, liquid oxygen, liquid argon, liquid helium, etc are not chemically pure. They normally consist impurities like air bubbles, suspended particles, etc. These impurities reduce the breakdown strength of these liquids considerably. When breakdown occurs in in these commercial liquids, additional gases and bubbles are evolved and solid decomposition products are formed. Due to this, electrode surface becomes rough and at times explosive sounds are heard due to generation of impulsive pressure through the liquid. Normally, breakdown in commercial liquids are classified as:

(i) Suspended particle mechanism

(ii) Cavitation and bubble mechanism
(iii) Stressed oil volume mechanism

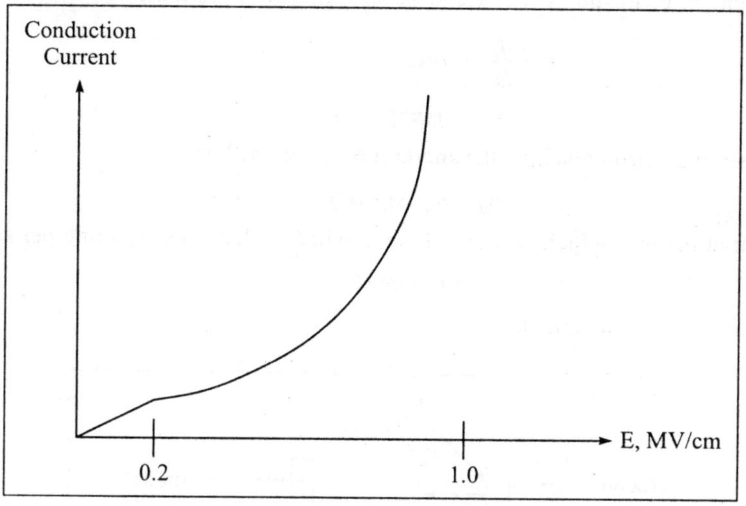

Fig. 3.3 Conduction current vs. electric field characteristic in hydrocarbon liquid.

3.3 Electrical Conduction in Gases

Gases are very easily available in the nature. Most of the gases are found everywhere in the nature, and so they are used widely in electrical engineering primarily for insulating purposes. When the applied voltage is low, a small current flows between the electrodes via gaseous medium separating these electrodes, and the gas retains its all insulating properties. But when the voltage applied is large enough, a sharp increase in currents between the electrodes through the gaseous medium is observed leading to electrical breakdown of the gaseous medium. A strongly conducting spark formed during breakdown produces a short circuit between the electrodes with heavy currents. The maximum voltage applied to the insulation (gas) at the moment of breakdown in called breakdown voltage.

A gas in its normal state is almost a perfect insulator. However, when a high voltage is applied between two electrodes immersed in a gaseous medium, the gas becomes a conductor and an electrical breakdown occurs. The processes primarily responsible for the breakdown of a gas are ionization by collision, photo-ionization and secondary emission processes.

In ionization process, an electron from a gas molecule is produced with simultaneous production of a positive ion. A free electron collides with a neutral gas molecule and gives rise to a new electron and a positive ion. In the figure shown below, if the energy E of the electron gained during travel between collisions exceeds the ionization potential V_i, which is the energy required to dislodge the electron from its atomic shell, then the ionization takes place.

$$e^- + A \xrightarrow{E > V_i} e^- + A^+ + e^- \tag{3.10}$$

where A is the atom, A^+ is the positive ion and e^- is the electron.

A few electrons produced by ultraviolet rays falling on the cathode, ionize neutral gas particles producing positive ions and additional electrons. These additional electrons make ionizing collisions and thus the process repeats itself. The number of electrons reaching the anode is greater than those liberated at the cathode. The positive ions also reach the cathode and give rise to secondary electrons.

In the figure 3.4, let n_0 be the number of electrons emitted from cathode and α (called Townsend's first ionization coefficient) be the average ionizing collisions made by electron per centimeter travel in the direction of the field, n_x be the number of electrons at a distance x from the cathode. At $x = 0$, $n_x = n_0$.

Similarly,
$$\frac{dn_x}{dx} = \alpha n_x$$

$$n_x = n_0 \exp(\alpha x) \tag{3.11}$$

Then the number of electron reaching the anode (i.e., $x = d$) will be

$$n_d = n_0 \exp(\alpha d) \tag{3.12}$$

And the current in the gap, which is equal to the number of electrons traveling per second is

$$I = I_0 \exp(\alpha d) \tag{3.13}$$

where I_0 is the initial current at cathode.

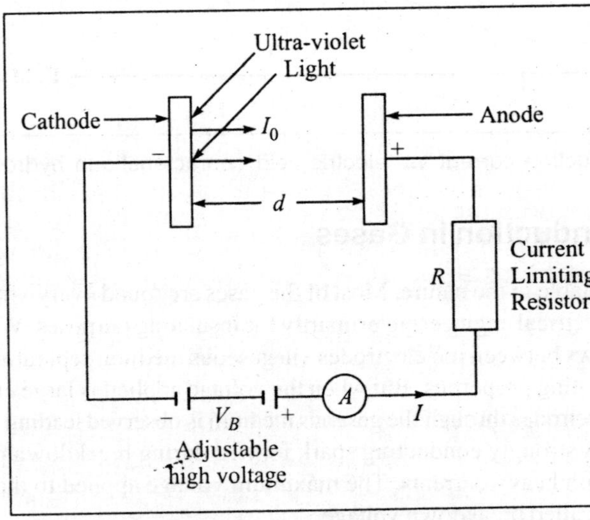

Fig. 3.4 Arrangement for Townsend discharge

3.4 Arc Discharge, Electric Breakdown in Gases

We mentioned how conduction in gases is the result of breakdown in gaseous medium due to various reasons. In the above-mentioned process, the probability of additional new electrons being liberated in the gap by other mechanisms increases and these new electrons further intensify the electron avalanche process. There are other mechanisms for liberation of electrons. The positive ions liberated may have sufficient energy to liberate electrons from the cathode when they impinge on it. The excited atoms or molecules in avalanche may emit photons, and this will lead to the emission of electrons due to photoemission. And, the metastable particles may diffuse back causing electron emission. The secondary electrons produced by these mechanisms are defined by secondary ionization coefficient γ. Let n_0' be the number of secondary electrons produced due to secondary processes and n_0'' be the total number of electrons leaving the cathode, then

$$n_0'' = n_0 + n_0' \tag{3.14}$$

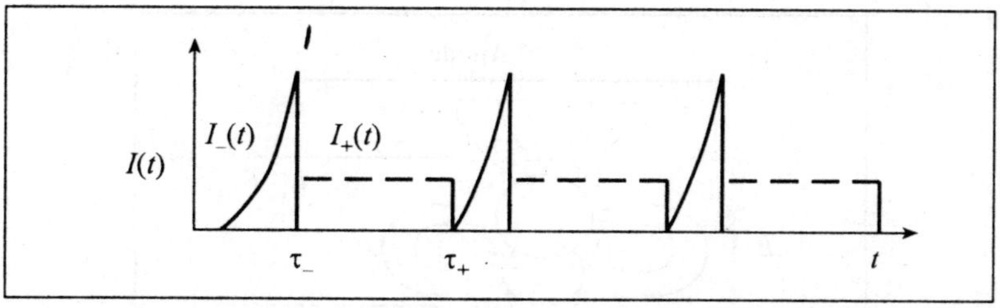

Fig. 3.5 Current as a function of time (Secondary electrons are produced at cathode by positive ions)

The total number of electrons reaching the anode now becomes

$$n = n_0'' \exp(\alpha d) = (n_0 + n_0') \exp(\alpha d) \quad (3.15)$$

And
$$n_0' = \gamma[n - (n_0 + n_0')] \quad (3.16)$$

So,
$$n = \frac{n_0 \exp(\alpha d)}{1 - \gamma[\exp(\alpha d) - 1]} \quad (3.17)$$

$$I = \frac{I_0 \exp(\alpha d)}{1 - \gamma[\exp(\alpha d) - 1]} \quad (3.18)$$

The coefficients α and γ can be calculated by plotting V-I characteristics for different gap settings. The initial slope of log (I_0/I) plot versus gap distance gives the value of α. By using the value of α, γ can be found by equation (3.18) using points on the upcurving portion of the graph.

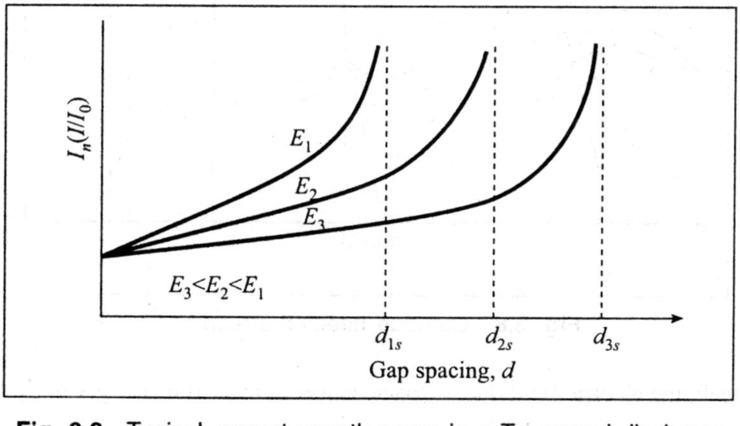

Fig. 3.6 Typical current growth curve in a Townsend discharge

In addition to Townsend mechanism of electric breakdown in gases, there is another theory known as streamer theory which is able to explain many nuances not explained by Townsend discharge mechanism. This theory predicts the development of spark discharge directly from a single electron avalanche in which the space charge developed by the avalanche itself is said to transform the avalanche into a plasma streamer. As shown in figure 3.7, a single electron starting at the cathode by ionization builds up an avalanche that crosses the gap. The electrons in the avalanche move very fast compared with the positive ions. So by the time electrons reach the anode, positive ions are at their initial position forming a positive

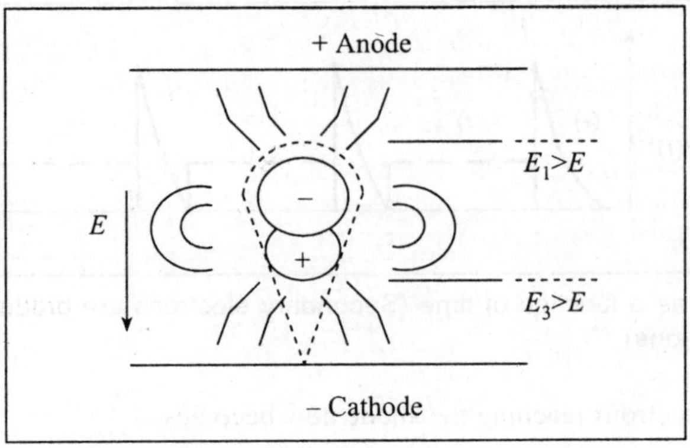

Fig. 3.7 Effect of space charge produced by an avalanche on the applied field

space charge layer at anode which enhances the field and secondary avalanche are formed due to presence of few electrons generated due to photo ionization in the space charge region. Although this occurs initially at the anode, due to the very high speed of the process spreads, space charge region will extend up to the cathode. This process creates a luminous track from anode to the cathode, which is called streamer. As soon as the streamer tip approaches the cathode, a cathode spot is formed and a stream of electrons rushes from the cathode to neutralize the positive charge in the streamer. And, the result is a spark, so spark breakdown has occurred.

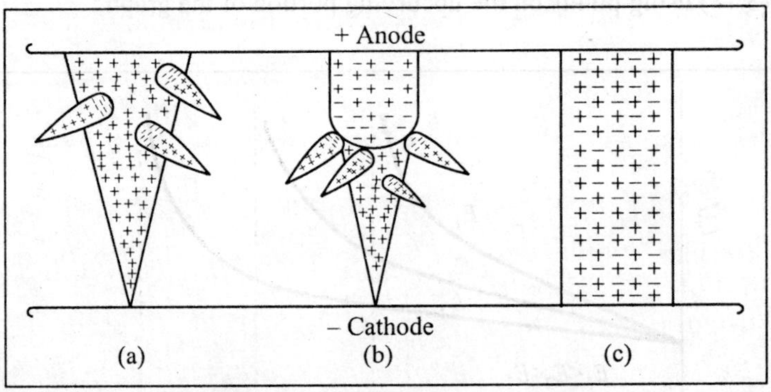

Fig. 3.8 Cathode directed streamer

According to Meek, the electric field due to space charge at the radius r is given by

$$E_r = 5.27 \times 10^{-7} \frac{\alpha \exp(\alpha x)}{(x/p)^{1/2}} \text{ V/cm} \tag{3.19}$$

where α is the Townsend's first ionization coefficient, p – gas pressure in torr, and x – distance to which the streamer has extended in the gap.. According to Meek the minimum breakdown voltage is obtained when $E_r = E$ and $x = d$ in the above equation. The equation then simplifies into

$$\alpha d + \ln(\alpha/p) = 14.5 + \ln(E/p) + 0.5 \ln(d/p) \tag{3.20}$$

The equation is solved between α/p and E/p at which a given p and d satisfy the equation. The breakdown voltage is given by the corresponding product of E and d. Based on the experimental results,

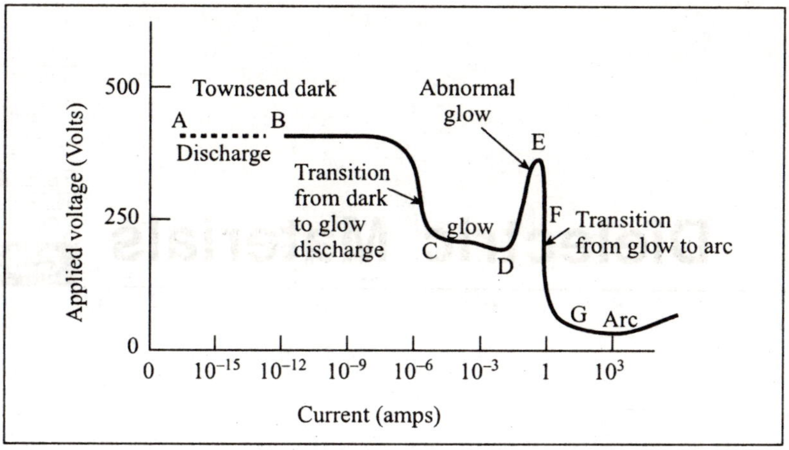

Fig. 3.9 *DC* voltage-current characteristics of an electrical discharge with electrodes having no sharp points or edges.

the breakdown potential of air is expressed as a function of product of pressure p and distance d as

$$V = 24.2 \times \frac{293pd}{760T} + 6.08 \times \left[\frac{293pd}{760T}\right]^{1/2} \tag{3.21}$$

where p is in torr, T in Kelvin and d in cm.

If the current in the gap is increased to about 1A or more, the voltage across the gap suddenly reduces to few volts ~ 20 – 25 V. The discharge becomes very luminous and noisy as shown in region *E-G* of figure 3.9. This phase is called arc discharge. The current density in the cathode region during the arc discharge increases to very high values ~ 10^3 to 10^7 A/cm^2 and the temperature also reaches to very high values ranging from 1000° C to several thousand degrees Celsius. The discharge contains very high electron density and positive ion density, which together are called arc plasma.

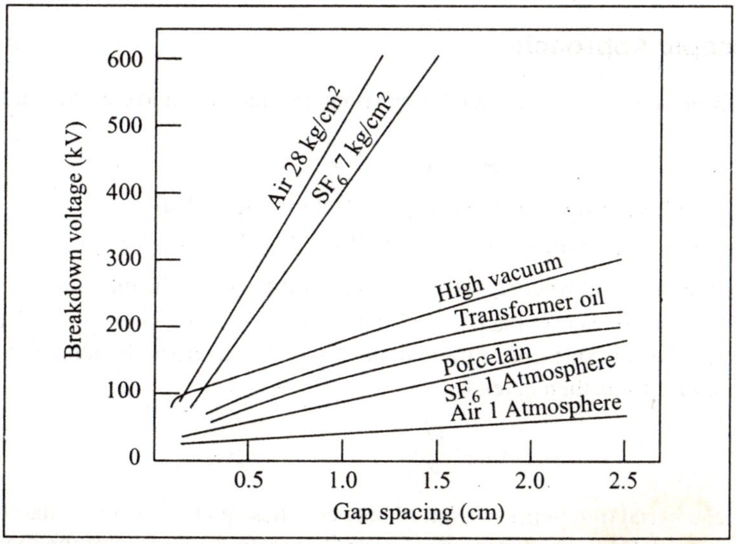

Fig. 3.10 *DC* breakdown strength of typical solid, liquid, gas and vacuum insulation in uniform fields.

4 Chapter

Dielectric Materials

4.1 Macroscopic Effects

4.1.1 Introduction

The importance of study of dielectric materials is justified by the fact that as the conductors are meant to conduct electricity with as low loss (low resistance) as possible, few also require these current carrying conductor at various high voltages to be insulated from any unwanted contact and this is achieved by using dielectric materials. As we know the dielectrics are the materials, which do not conduct electricity. There are very few free electrons in the conduction band at normal temperature and moderate electric field. The energy gap between valence band and conduction band is of the order of 5eV. The boundary between semiconductor and dielectric is assumed to be equal to 2.5eV in terms of energy gap. The failure of insulation is a big problem for various electric utilities as they use many generators, transformers and associated accessories whose normal functioning depends on the insulation used. The variation (increase) in the applied electric field can affect the normal functioning of insulating materials used for this or that application.

4.1.2 Macroscopic Approach

The relationship between the electric flux density at a point in a material and electric field strength is given by

$$D = \varepsilon E \tag{4.1}$$

Where D = Electric Flux density, Cm^{-2}, E = Electric field strength, Vm^{-1}

ε = Permittivity of the medium, $\varepsilon = \varepsilon_0 = 8.854 \times 10^{-12}\ Fm^{-1}$ for free space

And $\varepsilon = \varepsilon_0 \varepsilon_r$ for other material where ε_r is the relative permittivity of the material.

When a voltage V is applied across the parallel plate capacitor as shown in figure 4.1.a., charges $+Q_0$ and $-Q_0$ will be induced on the surfaces of these plates based on electrostatic induction. The capacitance of the parallel plate capacitor is then given by

$$C_0 = \frac{Q_0}{V_0} \tag{4.2}$$

When a dielectric material (medium) is inserted between these plates, then the charges on the surfaces of the plates get increased due to the appearance of the charge on the surface of the dielectric near to the capacitor plates.

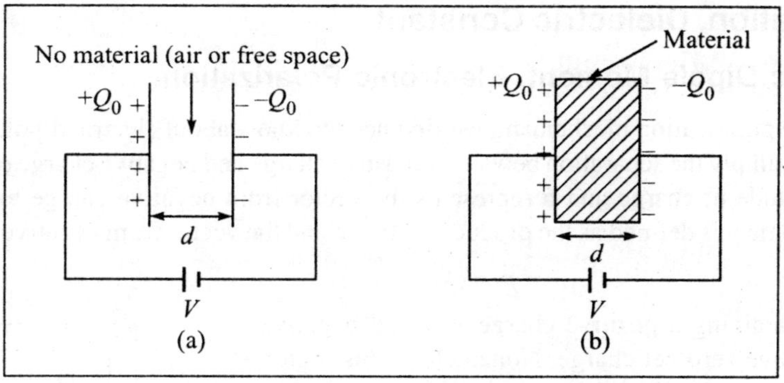

Fig. 4.1 Surface charges on metallic plates. Without material (a); with material (b).

In vacuum, the electric flux density is numerically equal to the surface charge, i.e., per unit area

$$\sigma_0 = Q_0 = \varepsilon_0 \frac{V}{d} \tag{4.3}$$

Where d is the separation between the plates.

Similarly, when there is a dielectric medium inserted between the plates, the above relationship becomes

$$\sigma = Q = \varepsilon_0 \varepsilon_r \frac{V}{d} \tag{4.4}$$

The relative permittivity can be defined as increase in charge stored on the capacitor plates due to insertion of the medium compared to the charge without medium.

$$\varepsilon_r = \frac{Q}{Q_0} = \frac{C}{C_0} \tag{4.5}$$

Increase in charge density when dielectric medium is inserted compared with that without dielectric is called polarization, i.e., $P = \sigma - \sigma_0 = Q - Q_0$

$$P = \varepsilon_0 \varepsilon_r \frac{V}{d} - \varepsilon_0 \frac{V}{d}$$

or, $$P = \varepsilon_0 \frac{V}{d} (\varepsilon_r - 1) \tag{4.6}$$

or, $$P = \varepsilon_0 (\varepsilon_r - 1) E \tag{4.7}$$

or, $$P = \varepsilon_0 \chi E \tag{4.8}$$

where $E = V/d$ electric field strength

$\chi = \varepsilon_r - 1 =$ dielectric susceptibility

$P =$ Polarization vector of the material

Polarization P can also be defined as the induced charge within dielectric per unit area.

$$P = \frac{Q'}{A} \tag{4.9}$$

From the above relationship we can conclude that for the value of relative permittivity equal to unity, there will be no polarization ($P = 0$), hence, no additional surface charge on the plate.

4.2 Polarization, Dielectric Constant

4.2.1 Electric Dipole Moment, Electronic Polarization

To understand the polarization mechanism, we first need to know about electric dipole moment. Electric dipole moment is simply the separation between a positive charge and negative charge of equal magnitude. If Q is the magnitude of charge and **a** represents the vector from negative charge to the positive, then electric dipole moment is defined as the product of charge and the vector from negative to positive charge.

$$p = Qa \tag{4.10}$$

In a region containing a positive charge $+Q$ and negative charge $-Q$ will have zero net charge. Nonetheless, this region will have net dipole moment if their charge centers do not coincide.

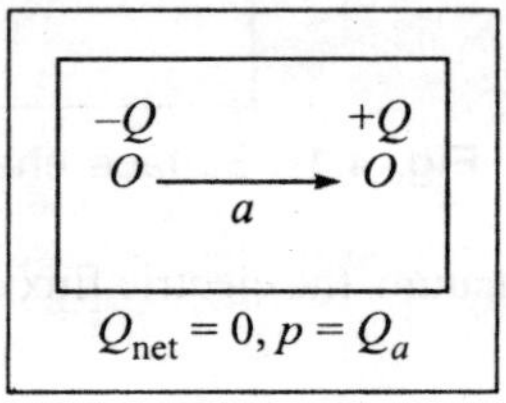

Fig. 4.2 Interpretation of electric dipole moment

In an atom when there is no applied field, the center of mass of atom and the charge center coincide with each other. So according to the definition of electric dipole moment, the dipole moment within the atom is zero. But when an external field is applied, the atomic center of mass does not change as the mass of an atom is mostly due to the contribution from the nucleus. In the other hand, the charge center for positive charge is unchanged; the negative charge center will be shifted in the direction opposite to the direction of applied field. The separation of charge centers means that there is induced electric dipole moment which is termed as polarization. Whenever an atom is said to be polarized, it possesses an effective dipole moment.

The induced dipole moment depends primarily on the electric field causing it. Induced dipole moment is related to applied electric field as

$$p_{ind} = \alpha E \tag{4.11}$$

where α= polarizability of the atom.

Since the polarization of a neutral atom involves the displacement of electrons, α is called electronic polarizability or coefficient of electronic polarization. Almost all atoms possess electronic polarizability as electrons are not rigidly fixed in atom.

Electronic polarizability can be calculated using the following expression (can be derived):

$$m_e \omega_0^2 r = \frac{Ze^2}{4\pi\varepsilon_0 r^2} \tag{4.12}$$

$$\alpha_e = \frac{Ze^2}{m_e \omega_0^2} \tag{4.13}$$

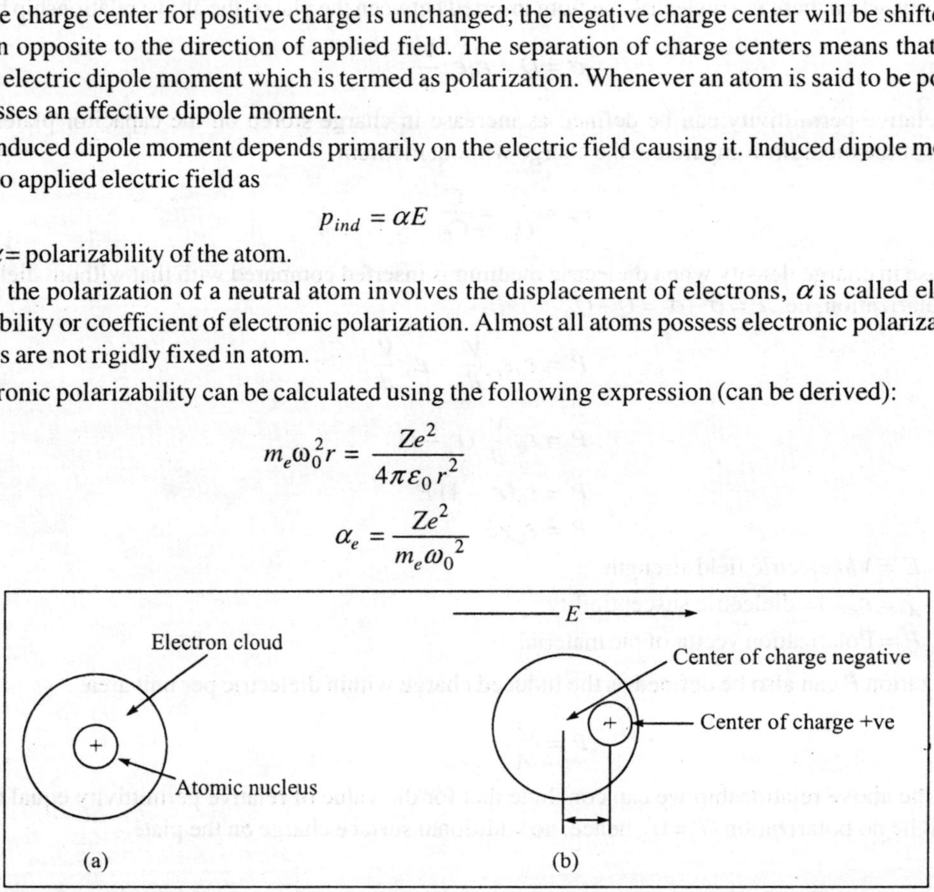

Fig. 4.3 A neutral atom $p = 0$ (a); induced dipole moment p ^{1} 0 (b)

Where Z = atomic number

ω_0 = oscillation frequency of the center of mass of the electron cloud about the nucleus,

m_e = mass of electron and, e = charge of electron

4.2.2 Polarization Vector and Local Electric Field

When a material is placed in an electric field, the atoms and the molecules of the material become polarized and the dipole moments are distributed in the material as shown in figure 4.4. If we consider only the bulk (fig 4.4.b), the dipoles are aligned head to tail. Every positive charge has a negative charge next to it and vice versa. There is no net charge within the material. But the negative charges of dipoles appearing on the left hand side of the material are not cancelled by positive charge of any dipoles at this face and in the same manner the positive charges of the dipole appearing on the right hand side are also not cancelled by any dipoles. The negative charge and positive charge on the far sides of the material are bound to the surface charge on the plate. So there are positive and negative surface charges on the right side and left side of the material represented here by $+Q_p$ and $-Q_p$ respectively. These charges shown in figure 4.4.c, which are bound, are called surface polarization charges.

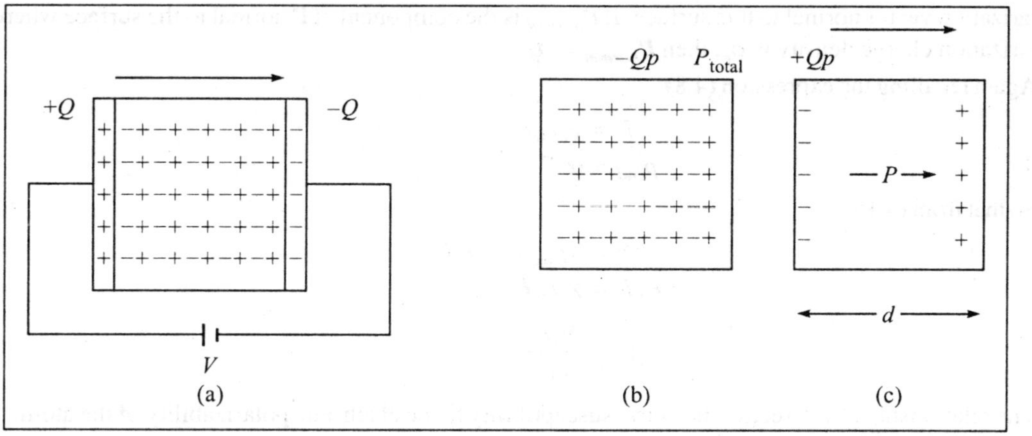

Fig. 4.4 Polarization of dielectric in the influence of applied field (a); Bound molecular polarization (b); Surface polarization (c).

Polarization of a dielectric medium is represented by polarization vector **P** and defined as dipole moment per unit volume.

$$P = \frac{\sum_{1}^{N} p_i}{V} \tag{4.14}$$

where p_i = dipole moment of ith molecule,

N = number of molecules within the material,

V = volume of the dielectric material.

If p_{av} is the average dipole moment per molecule in the dielectric, then an equivalent P is given by

$$P = Np_{av} \tag{4.15}$$

If $+Q_p$ and $-Q_p$ are the surface polarization charges and d the separation between them, then total dipole moment within the material is

$$P_{total} = Q_p d \tag{4.16}$$

Now the polarization is given by

$$P = \frac{P_{total}}{V}$$

$$P = \frac{Q_p d}{Ad} \tag{4.17}$$

$$P = \frac{Q_p}{A}$$

But as mentioned in equation (4.9), Q_p/A is also the surface polarization charge density. So

$$P = \sigma_p \tag{4.18}$$

Where σ_p = surface polarization charge density.

The charge per unit area appearing on the surface of a polarized medium is equal to the component of the polarization vector normal to this surface. If $\boldsymbol{P}_{normal}$ is the component of **P** normal to the surface where the polarization charge density is σ_p, then $\boldsymbol{P}_{normal} = \sigma_p$.

Again recalling the expression (4.8)

$$P = \chi_e \varepsilon_0 E$$

and

$$p_{ind} = \alpha_e E$$

So that from (4.15)

$$P = Np_{ind} = N\alpha_e E \tag{4.19}$$

$$\chi_e \varepsilon_0 E = \chi_e \varepsilon_0 E$$

or,

$$\chi_e = \frac{N\alpha_e}{\varepsilon_0} \tag{4.20}$$

The relationship (4.20) relates dielectric susceptibility to the electronic polarizability of the atom.

The relative permittivity is related to dielectric susceptibility as

$$\varepsilon_r = 1 + \chi_e$$

$$\varepsilon_r = 1 + \frac{N\alpha_e}{\varepsilon_0} \tag{4.21}$$

The expression (4.21) relates macroscopic property of the material that is relative permittivity to the microscopic property that is electronic polarizability.

The actual field experienced by a molecule in dielectric is defined as local field $\boldsymbol{E}_{loc}$ which depends not only on the free charges on the plates but also on the arrangement of all polarized molecules around this point. The greater the polarization the greater is the local field. In a material with cubic structure or a liquid, the local field is given by

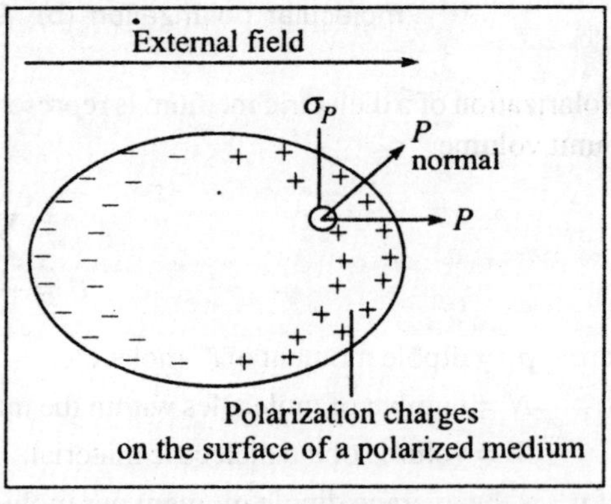

Fig. 4.5 Polarization charge density on the surface of polarized medium.

$$E_{loc} = E + \frac{P}{3\varepsilon_0} \tag{4.22}$$

Induced polarization of molecule depends on this local field rather than applied field E.

Thus,
$$p_{ind} = \alpha_e E_{loc} \tag{4.23}$$

The fundamental definition of susceptibility given by equation (4.8) is unchanged.

$$\therefore \quad P = Np_{ind} = N\alpha_e E_{loc} = N\alpha_e\left(E + \frac{P}{3\varepsilon_0}\right)$$

or,
$$P = N\alpha_e E + \frac{N\alpha_e P}{3\varepsilon_0}$$

or,
$$P = \frac{N\alpha_e E}{1 - \frac{N\alpha_e}{3\varepsilon_0}}$$

But from (4.8), we have

$$P = \chi_e \varepsilon_0 E.$$

$$\chi_e \varepsilon_0 E = \frac{N\alpha_e E}{1 - \frac{N\alpha_e}{3\varepsilon_0}}$$

$$N\alpha_e = \chi_e \varepsilon_0 \left(1 - \frac{N\alpha_e}{3\varepsilon_0}\right)$$

$$N\alpha_e = \chi_e \left(\frac{3\varepsilon_0 - N\alpha_e}{3}\right) = \chi_e \varepsilon_0 - \chi_e \frac{N\alpha_e}{3}$$

$$N\alpha_e + \chi_e \frac{N\alpha_e}{3} = \chi_e \varepsilon$$

$$N\alpha_e \left(1 + \frac{\chi_e}{3}\right) = \chi_e \varepsilon_0$$

$$N\alpha_e \left(\frac{3 + \chi_e}{3}\right) = \chi_e \varepsilon_0$$

$$\frac{N\alpha_e}{3\varepsilon_0} = \frac{\chi_e}{3 + \chi_e} = \frac{\varepsilon_r - 1}{\varepsilon_r + 2} \tag{4.24}$$

And, at optical frequencies, ε_r is replaced by n^2, where n is refractive index.

Equation (4.24) is known is Clausius-Massotti equation, which relates relative permittivity with electronic polarizability more accurately. There are different polarization mechanisms in materials. The different polarization mechanisms are classifieds in to:

(i) Electronic polarization
(ii) Ionic polarization

(iii) Orientational (dipolar) polarization
(iv) Interfacial polarization

Of the above-mentioned, electronic polarization has been discussed already. So we will be discussing other remaining three types of polarization mechanisms.

4.2.3 Ionic Polarization

Ionic crystals like NaCl, KCl, LiBr, etc witness ionic polarization. The ionic crystals have distinctly distinguishable ions located at well-defined lattice sites so that each pair of oppositely charged neighboring ions has a dipole moment. As shown in the figure 4.6.a, in the absence of applied field, there is no net polarization of the specimen because the dipole moments of equal magnitude are lined up head to head and tail to tail. The dipole moment p_+ in the $+ve$ x direction has the same magnitude as dipole moment p_- in the $-ve$ x direction giving zero net polarization. When external electric field is applied, Cl^- ions are pushed opposite to the direction of applied field and Na^+ ions in the direction of field about their equilibrium positions. Consequently, dipole moment p_+ in the direction of $+x$ increases to p'_+ and that in the $-x$ direction decreases to p'_- and the net dipole moment is no longer zero. The average dipole moment per ion pair $(p'_+ - p'_-)$ in the presence of electric field depends on the field. The ionic polarizability α_i is defined in terms of local field experienced by the ions as

$$p_{av} = \alpha_i E_{loc} \tag{4.25}$$

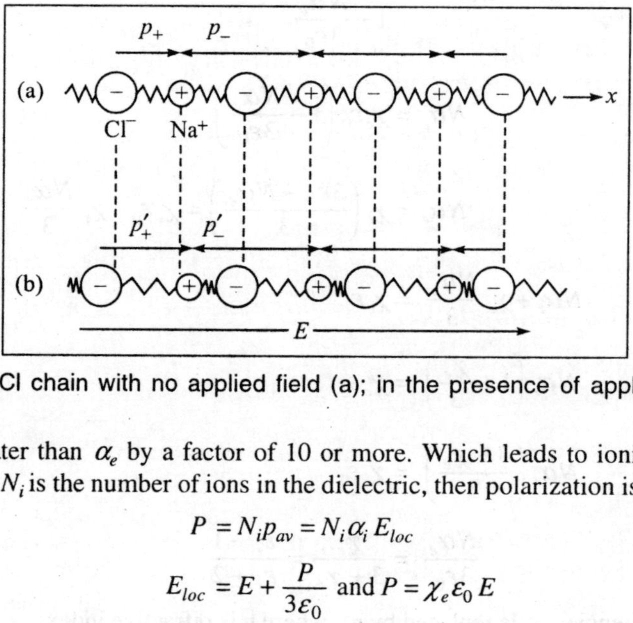

Fig. 4.6 NaCl chain with no applied field (a); in the presence of applied field (b).

Generally α_i is greater than α_e by a factor of 10 or more. Which leads to ionic solids having large dielectric constants. If N_i is the number of ions in the dielectric, then polarization is given by

$$P = N_i p_{av} = N_i \alpha_i E_{loc} \tag{4.26}$$

But
$$E_{loc} = E + \frac{P}{3\varepsilon_0} \text{ and } P = \chi_e \varepsilon_0 E$$

After some manipulations as in the previous case, we can obtain the familiar Clausius-Massotti equation for ionic polarization.

$$\frac{\varepsilon_r - 1}{\varepsilon_r + 2} = \frac{1}{3\varepsilon_0} N_i \alpha_i \tag{4.27}$$

Electronic polarization also persists in these solids due to the displacement of negative charge center by the applied field. But in ionic solids their contribution is much smaller and can be neglected.

4.2.4 Orientational (Dipolar) Polarization

Certain molecules like HCl possess permanent dipole moments. However, in the absence of external electric field, these dipole moments are randomly oriented due to thermal agitation and the average dipole moment per molecule in the sample is zero. In HCl the dipole moment denoted by p_0 is directed from Cl^- ion to H^+ ion giving. When an external field E is applied, both the ions experience force in opposite directions. But the nearly rigid bond between these ions holds them together. The molecule thus experiences a torque τ about its center of mass, which tries to rotate dipole p_0 and align with the applied field. If all the molecules were to simply rotate and align with the field, the polarization of the solid would be

$$P = Np_0 \tag{4.28}$$

where N = number of molecules per unit volume,

P = Polarization vector for the specimen.

However, due to the thermal energy , the molecules collide with each other and with the container there by destroying some of the dipole alignments which means the orientations of dipole moments are randomized. Nonetheless, there is net polarization of the sample which would have been greater had there not been the random orientation of the molecules. This type of polarization is termed as Orientational (dipolar) polarization.

When the dipole will be aligned by the external field, certain amount of work will be done. No work will be done if the dipole is already aligned with the field. Maximum work will be done when the torque has to rotate the field by an angle of 180°. The torque experienced by dipole is

$$\tau = (F \sin \theta)a = (EQ \sin \theta)a = p_0 E \sin \theta \tag{4.29}$$

If we take potential energy equat to zero for $\theta = 0$, then the its maximum would be for the angle $\theta = 180°$, so

$$PE_{max} = \int_0^{180°} (p_0 E \sin \theta) d\theta = p_0 E \tag{4.30}$$

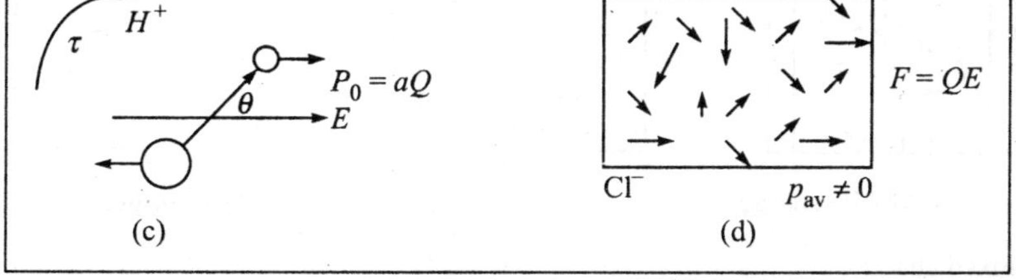

Fig. 4.7 HCl molecule possesses permanent dipole moment (a); Net average dipole moment is zero in the absence of external field (b); Permanent dipole experiencing a torque due to applied field (c); Presence of net average dipole moment per molecule due to the external applied field.

The average dipole potential energy is $\mathbf{PE}_{max}/2$. The average dipole energy along applied field is directly proportional to permanent dipole moment and the ratio of average dipole energy to the average thermal energy of the dipole gives the average dipole moment per molecule

$$p_{av} \propto p_0 \frac{p_0 E}{5kT/2} \tag{4.31}$$

where $5kT/2$ is the average thermal energy per molecule (for 5 degrees of freedom).

But by using Boltzmann statistics (see Appendix-2), we would find the average dipole moment per molecule as given by

$$p_{av} = \frac{p_0^2 E}{3kT} \tag{4.32}$$

As we already defined dipole moment as proportional to the applied field, so from equation (4.32), we can introduce the proportionality constant termed as dipolar or orientational polarizability and given by

$$\alpha_d = \frac{p_0^2}{3kT} \tag{4.33}$$

Dipolar (orientational) polarization is strongly temperature dependent and decreases with increase in temperature and so is the relative permittivity. Polar liquids such as water, alcohol and various electrolytes and polar gases such as gaseous HCl and steam exhibit this type of polarization.

4.2.5 Interfacial Polarization

Whenever there is accumulation of charge at an interface between two materials or two regions, there is interfacial polarization. The accumulation of charge near the dielectric in one of the electrodes takes place as shown in figure. There are always some defects and impurities within the dielectric material, which contribute, free charge carriers such as electron and holes. In a dielectric material like ceramic or glass, there are always equal numbers of positive and negative ions but positives ions like H^+ and Li^+ are far more mobile than the negative ions in the structure. Under the influence of the applied field, these positive ions migrate near to the negative electrode. These positive ions cannot leave the dielectric material and enter the crystal structure of metallic electrode. So, these positive ions simply pile up at the interface and give

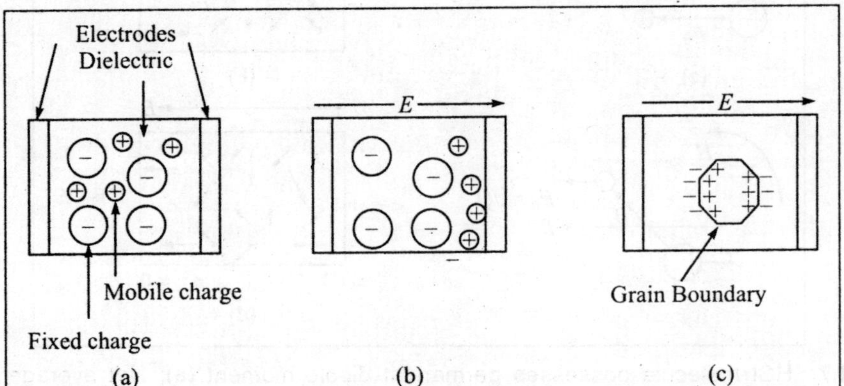

Fig. 4.8 No polarization (a); Net polarization with the field (b) and (c)

rise to a positive space charge near the negative electrode. These positive charges at the interface attract more electrons to the negative electrode and this additional charge on the electrode appears as an increase in the dielectric constant of the material. The positive charge near the interface with the electrode and negative charge that remain in the bulk together constitute dipole moments that appear in the polarization vector P and so the polarization due to this is termed as interfacial polarization.

Trapping of electrons or holes at the crystal defects near the interface between electrode and the dielectric is another example of interfacial polarization. The mobile charge carriers drifting under the influence of external field try to reach the electrodes but are trapped at these defects. Grain boundaries also lead to interfacial polarization, as the mobile charges cannot move beyond the grain boundaries, so give rise to the polarization.

The induced dipole moment per molecule in the presence of electronic, ionic and orientational polarizations is given by

$$p_{av} = \alpha_e E_{loc} + \alpha_i E_{loc} + \alpha_d E_{loc} \tag{4.34}$$

Each effect simply adds to the net dipole moment per molecule the fact that is verified by experiments. But the interfacial polarization cannot be simply added to the above because it occurs at the interfaces and cannot be put into an average dipole moment per molecule in the bulk. The local fields also cannot be simply approximated for Orientational polarization. So the dielectric constant under electronic and ionic polarization mechanisms can be found from

$$\frac{\varepsilon_r - 1}{\varepsilon_r + 2} = \frac{N_e \alpha_e + N_i \alpha_i}{3\varepsilon_0} \tag{4.35}$$

Table 4.1 Typical examples of polarization mechanisms

Polarization	*Example*	*Static* ε_r	*Comments*
Electronic	Ar gas	1.0005	
Electronic	Ar liquid (T < 87.3 K)	1.53	
Electronic	Si crystal	11.9	Due to valence electrons, bond polarization
Ionic	NaCl crystal	5.9	
Ionic	CsCl crystal	7.2	
Dipolar	Water	80	
Dipolar	PVC	7	Partly hindered in solid

4.3 Dielectric Losses, Frequency and Temperature Effect

The static dielectric constant is the effect of polarization under dc conditions. When the applied field is sinusoidal ac, polarization of material under these conditions leads to the ac dielectric constant, which is generally different from the dc. As an example we will consider the Orientational polarization in dipolar liquids where the varying field tries to change the alignment of dipoles in one and the other way. If the instantaneously induced dipole moment per molecule can readily follow the field variations, then at any instant the dipole is

$$p = \alpha_d E \tag{4.36}$$

where the orientational polarizability has its expected maximum value from the dc conditions, that is,

$$\alpha_d = p_0^2/3kT \tag{4.37}$$

There are two factors opposing the immediate alignment of dipoles with the applied field. First, the thermal agitation tries to randomize the dipole orientations. Second, the molecules rotate in the viscous medium by virtue of their interactions with neighbors, which is particularly strong in liquids and solids. So the dipoles cannot respond instantaneously to the changes in the filed. At low frequencies the dipoles can respond with the change in applied field, so dipolar polarizability is maximum in this region. In high frequency range, the dipoles cannot respond to the change in applied field, consequently the high frequency field cannot polarize the material means it the orientational polarizability is zero.

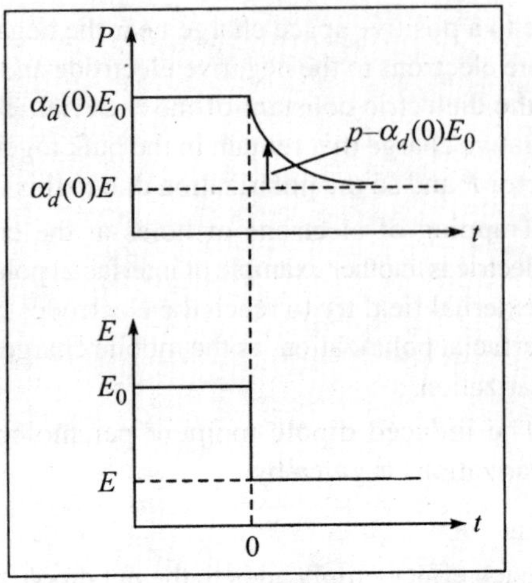

Fig. 4.9 When dc field is suddenly changed from E_0 to E at $t = 0$, the dipole moment p changes from $\alpha_d(0)\ E_0$ to $\alpha_d(0)\ E$

In the figure above, the dielectric is applied by a dc field E_0 for long time before the field is reduced to $E \cdot \alpha_d(0)$ corresponds to polarizability at zero frequency. The induced dipole moments are also reduced as shown in figure 4.9. The decrease in dipole moments is achieved by random collision between molecules. Assume τ as average relaxation time between molecular collisions. If p is the induced dipole moment, then $p - \alpha_d(0)\ E_0$ is the excess moment which must disappear as t tends to infinity. τ is the average time to eliminate the access dipole moment. Rate of change of dipole moment is

$$\frac{dp}{dt} = -\frac{p - \alpha_d(0)E}{\tau} \tag{4.38}$$

The above equation can be used to describe dipolar polarizability in ac conditions. Electric field in ac condition is given by

$$E = E_0 \sin \omega t = E_0 \exp(j\omega t) \tag{4.39}$$

From equation (4.38) and (4.39), we have

$$\frac{dp}{dt} = -\frac{p}{\tau} + \frac{\alpha_d(0)}{\tau} E_0 \exp(j\omega t)$$

$$p = \alpha_0 E_0 \exp(j\omega t) \tag{4.40}$$

where
$$\alpha_0(\omega) = \frac{\alpha_d(0)}{1 + j\omega\tau} \tag{4.41}$$

$\alpha_d(\omega)$ represents dipolar orientational polarizability under ac conditions. Equation (4.40) indicates that dipole moments and applied field are out of phase. So is the polarization vector $P = Np$ and E. At low frequencies, when $\omega\tau << 1$, $\alpha_d(\omega)$ is nearly equal to $\alpha_d(0)$, and dipole moment p is in phase with the applied field., the rate of relaxation, $1/\tau$, is much faster than the frequency of applied field or the rate at which the polarization is being changed, so p closely follows E_0. At higher frequencies i.e. $\omega\tau >> 1$, the rate of relaxation is very slow than the frequency of applied field and p can no longer follow E, so the polarization is zero.

From the complex dipolar polarizability, we can obtain the complex dielectric constant as polarizability is related to dielectric constant. So, the dielectric constant in ac is written as

$$\varepsilon_r = \varepsilon_r' - j\varepsilon_r'' \tag{4.42}$$

where $\varepsilon_r' = \varepsilon_{r\infty} + \dfrac{\varepsilon_r'(0) - \varepsilon_{r\infty}'}{1+\omega^2\tau^2}$ and $\varepsilon_r'' = \dfrac{\varepsilon_r'(0) - \varepsilon_{r\infty}'}{1+\omega^2\tau^2}$

where ε_r' is the real part and ε_r'' is the imaginary part, and both are frequency dependent.

$\varepsilon_{r\infty}'$ and $\varepsilon_r'(0)$ is dielectric constant at high and low frequencies. As shown in the figure 4.10, the real part decreases from maximum value $\varepsilon_r'(0)$ at dc corresponding to $\alpha_d(0)$, to unity at high frequencies when $\alpha_d(\omega) = 0$ as ω tends to infinity. The imaginary part ε_r'' is zero at low and high frequencies and peaks when $\omega\tau = 1$ or $\omega = 1/\tau$. The real part of the capacitance will be used to calculate the capacitance whereas the imaginary part will be used to calculate the energy loss in the dielectric under ac conditions. The energy loss is due to the random collision of dipoles under varying fields.

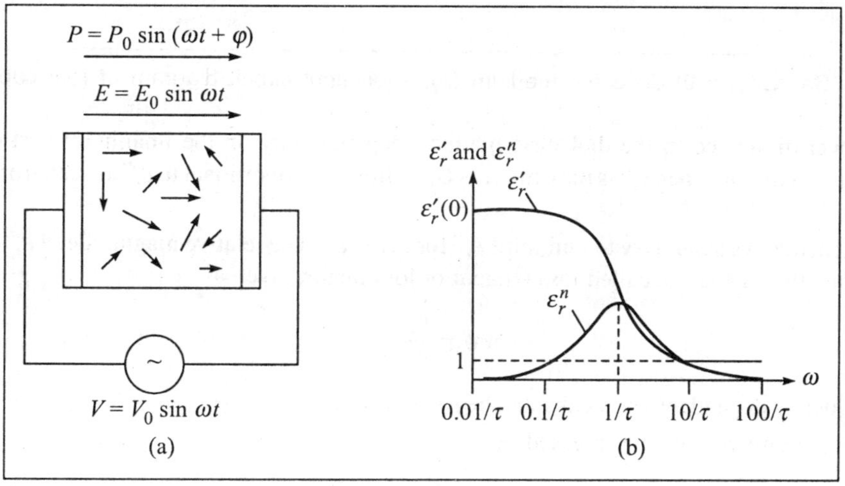

Fig. 4.10 *AC* field applied to dipolar medium (a). Relative permittivity in *ac* (b)

For a capacitor shown in figure 4.11.a, the admittance can be calculated as

$$Y = \frac{j\omega A\varepsilon_0\varepsilon_r(\omega)}{d} = \frac{j\omega A\varepsilon_0\varepsilon_r'(\omega)}{d} + \frac{\omega A\varepsilon_0\varepsilon_r''(\omega)}{d} \tag{4.43}$$

or,

$$Y = j\omega C + G_p \tag{4.44}$$

where

$$C = \frac{A\varepsilon_0\varepsilon_r'(\omega)}{d} \tag{4.45}$$

$$G_p = \frac{\omega A\varepsilon_0\varepsilon_r''(\omega)}{d} \tag{4.46}$$

C represents the capacitance of the capacitor and depends on the real part of dielectric constant, and G_p is called conductance of the capacitor and represents the power loss in the capacitor under varying electric field. The capacitor can be represented by an ideal capacitor with capacitance C and a resistor with resistance $R = 1/G_p$ in parallel to represent the power loss in the dielectric medium.

Here the there is no real power dissipation in C but there is real power dissipation in $G_p(R_p)$. The input power is given by

$$p_{in} = VI = V(VY) = j\omega CV^2 + V^2G_p \tag{4.47}$$

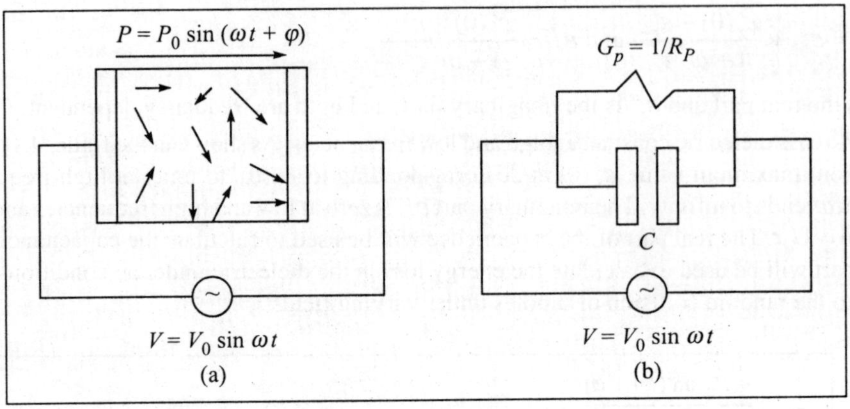

Fig. 4.11 Capacitor with dielectric medium (a), equivalent circuit diagram of real capacitor (b)

So the power dissipated in the dielectric medium depends only on the imaginary part of dielectric constant and is maximum when ε_r'' is maximum as G_p is directly proportional to ε_r'' as seen from expression (4.46).

For any capacitor, we should try to minimize ε_r'' for given ε_r'. The relative magnitude of ε_r'' with respect to ε_r' is defined through tan δ, called loss tangent or loss factor, given as

$$\tan \delta = \frac{\varepsilon_r''}{\varepsilon_r'} \tag{4.48}$$

which is frequency dependent and peaks just beyond $\omega = 1/\tau$.

Power loss per unit volume is expressed as

$$P_{vol} = \frac{V^2 G_p}{V_{olume}} = V^2 \frac{\omega A \varepsilon_0 \varepsilon_r''}{d} \frac{1}{dA}$$

$$P_{vol} = \frac{V^2}{d^2} \omega \varepsilon_0 \varepsilon_r'' = E^2 \omega \varepsilon_0 \varepsilon_r'' = E^2 \omega \varepsilon_0 \varepsilon_r' \tan \delta \tag{4.49}$$

From equation (4.49), it is clear that power loss per unit volume is the function of frequency, applied field and loss tangent.

We have obtained the frequency dependence of dielectric loss for dipolar orientational polarization. But this can be extended to other forms of polarization as well. The frequency dependence of dielectric loss is shown in figure 4.12 below.

The dependence of dielectric constant on temperature can be seen from expressions relating polarizability to the temperature. In most of the cases, increased temperature indicates toward increased thermal agitation of molecules, which always hinders the orientation of dipoles within the material. So the dielectric constant is smaller at higher temperature compared to that at lower temperature. The dielectric strength will be lower at higher temperature than at lower one.

4.4 Dielectric Breakdown

A defining property of dielectric material is not only its ability to increase capacitance but also and equally important is its insulating behavior. Dielectric materials are widely used as insulating media between

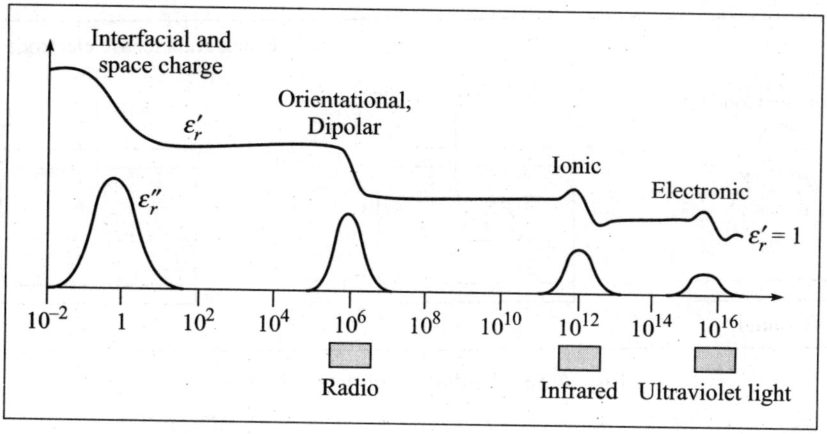

Fig. 4.12 Frequency dependence of real and imaginary parts of dielectric constant.

conductors at different voltages to prevent ionization of air and current flashovers between conductors. The voltage across conductors and hence within the dielectric cannot be increased without limit. A voltage will be eventually reached when substantial current starts flowing through the dielectric, which may lead to what is called ***dielectric breakdown***. In gases and liquid, the breakdown is not permanent but in solids, there can be permanent damage due to the breakdown. The maximum electric field that can be applied to the insulating medium without causing dielectric breakdown is called ***dielectric strength*** of the material and is denoted by $\boldsymbol{E_{br}}$. The dielectric strength of solid depends on molecular structure, impurities in the material, sample geometry, nature of the electrodes, temperature, ambient conditions, duration and frequency of applied field, and is different in ac and dc conditions. The ageing also affects the dielectric strength. Dielectric breakdown in solids, liquids and gases will be discussed separately.

4.4.1 Dielectric Breakdown in Gases

Due to cosmic radiation, there are always a few free electrons in a gas. If the field is sufficiently large, then one of these electrons can be accelerated to sufficiently large kinetic energies to impact ionize a neutral gas molecule and produce additional free electron and a positively charged gas ion. Both the first and liberated electrons are available to accelerate in the field and again impact-ionize more neutral gas molecules, and so on. The breakdown in gases depends upon pressure. Higher pressure means greater concentration of molecules and the reduced mean separation between molecules, hence the shorter mean free path of free electrons. This means free electrons should be accelerated by larger energies to impact ionize the neutral gas molecule. So, E_{br} is greater at higher pressure.

When a local region only is exhibiting discharge so that the discharge does not directly connect two electrodes, it is said that a partial discharge has occurred. For a high voltage cylindrical conductor above a grounded electrode, the electric field is highest near the surface of the conductor facing ground. This field initiated discharge locally because the field is sufficiently strong to continue the electron avalanche. But away from the conductor due to weak field, there will be no electron avalanche. Such local discharge is termed as corona discharge. Voids and cracks within the solid dielectric material and discontinuities at dielectric-electrode interface can also lead to partial discharge since the electric field in these voids and cracks are stronger than in dielectric material itself. Partial discharges in these voids and cracks erode the surrounding region of the dielectric both physically and chemically, which eventually can lead to major breakdown if not controlled in proper time.

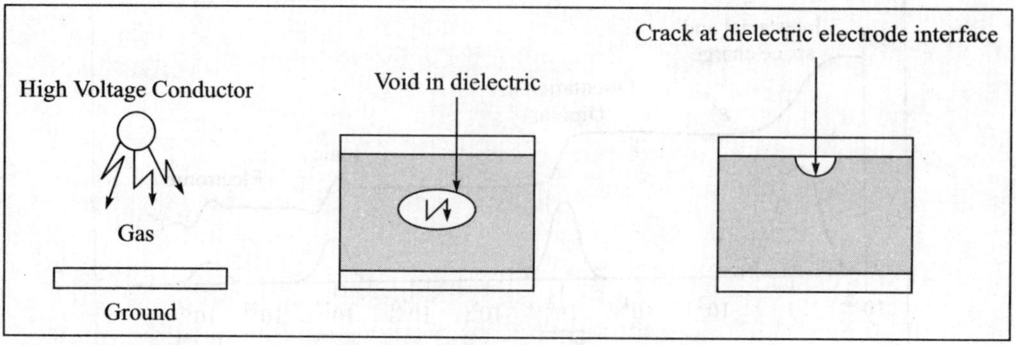

Fig. 4.13 Partial discharges in gases

4.4.2 Dielectric Breakdown in Liquids

The process that leads to the breakdown in liquid dielectric is not as clear as that in gases. In impure liquids with small particles in suspension, it is believed that these impurities coalesce end to end to form a conducting bridge between electrodes thereby give rise to breakdown. Sometimes discharge initiates as partial discharge in gas bubbles entrapped in the liquids. Moisture absorption and gas absorption deteriorates the dielectric strength of liquids. In oil in a long run, oxidation produces acidic products and some parts may become more conducting by this and can give rise to discharge. Electron emission from electrodes at high field also can initiate discharge in liquids.

4.4.3 Dielectric Breakdown in Solids

There are various mechanisms of dielectric breakdown in solids such as intrinsic breakdown, thermal breakdown, breakdown due to insulation aging, electromechanical breakdown, etc.

In intrinsic breakdown, which is also called electronic breakdown, an electron in the conduction band of dielectric can be accelerated to sufficiently large energy to impact ionize host atom. The collision of electron with energy more than the band gap energy with lattice vibrations is capable of exciting electron from valence band to conduction band, and, then further, both the primary and secondary electrons will impact ionize other host atoms thereby generating an electron avalanche effect. SiO_2 films with no micro-structural defects in MOS capacitors exhibits intrinsic breakdown. Although the theoretical limit for $E_{br} \sim$ 1MV cm^{-1}, due to micro-structural defects in reality there are lower values for breakdown.

Finite conductivity of the insulation means there is a Joule heat σE^2 being released within the solid. At high frequencies, dielectric loss $V^2 \omega \tan \delta$ is high. If the heat generated in the solid cannot be removed quickly by thermal conduction or other means, the temperature of dielectric increases and leads to higher electrical conductivity, which will cause more Joule heating. Certain local area can become hot spots and if there are many such hot spots in the sample there can be permanent conducting channel connecting two electrodes and hence a thermal breakdown of dielectric. At the same temperature the breakdown field can be drastically reduced if the duration is increased.

The basis of electromechanical breakdown is the force experienced by dielectric due to oppositely charged plates between which it is placed. The compressive force increases with the increase in voltage (hence the electric field). The thickness of the dielectric is reduced which will increase the charge on the electrode as given by $Q = CV = \varepsilon_0 \varepsilon_r AV/d$. This will lead to further increase in compressive force and decrease in thickness. This process will continue till the dielectric deforms plastically causing the insulation

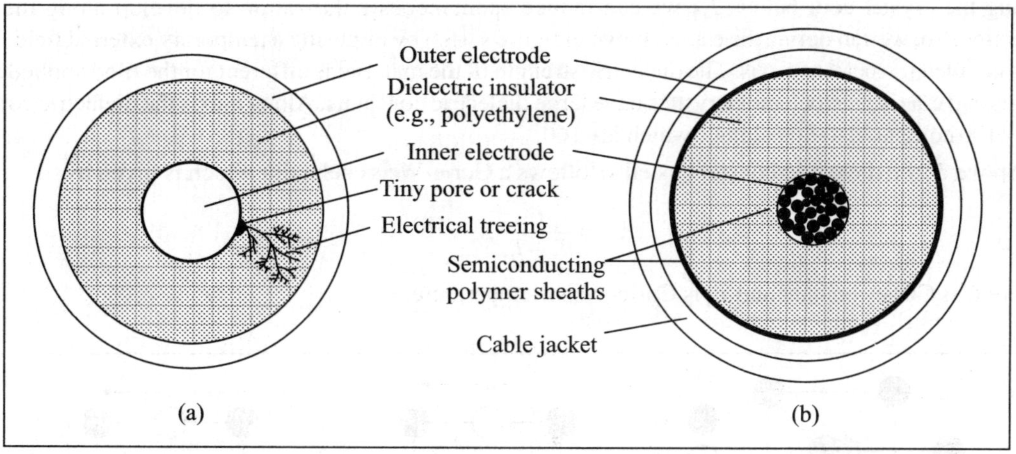

Fig. 4.14 Electrical treeing (a), high voltage coaxial cable with semiconductor sheath (b)

to breakdown by increase in field. The increase in E and decrease in d produces more Joule heat and dielectric loss heat causing more problems for mechanical stability of material. It is possible to reach thermal breakdown due to mechanical deformation. There is possibility of internal cracks being developed and these cracks experience shear forces, which combined with large electric field, can cause crack propagation and eventually mechanical and hence dielectric failure.

Internal discharges are the partial discharges that take place in micro-structural voids, cracks, or pores within dielectric. Initially the pores size can be small and the partial discharges insignificant, but with time partial discharges erode the internal surface of the void by locally melting the insulator. And, eventually an electrical tree type discharge is developed. The erosion of dielectric by partial discharge propagates like a branching tree, which are erosion channels in which gaseous discharge takes place and forms a conducting channel. In high voltage cables, to avoid this problem, the electrode is sheathed by an intermediate semiconducting material.

With the time of use the properties of insulating materials get degraded. And eventually dielectric breakdown takes place at fields below the values established by experiments. Aging determines the useful life of the insulating material. Although there are many factors which directly or indirectly affect the properties and performance of insulating material, we here are limited just to the fact that with time these materials loose their useful properties and the breakdown is eventual outcome.

4.5 Ferroelectricity and Piezoelectricity

4.5.1 Ferroelectricity

Certain crystals like Barium titanate ($BaTiO_3$) which have permanent polarization even in the absence of an applied field are called ferroelectric crystals and the phenomenon ferroelectricity. At above 130°C, the crystal structure of $BaTiO_3$ has cubic structure and the center of mass of positive and negative charges coincide and the net polarization is zero. Below 130°, $BaTiO_3$ has tetragonal structure and center of mass of positive and negative charges do not coincide. So there is polarization vector P as there is separation between negative and positive charge centers. The temperature above which the ferroelectricity is lost is called Curie temperature T_c. Below this the whole specimen becomes spontaneously polarized and above T_C, disorderedness of dipole increases which is called pyroelectric state. By applying temporary field and letting the crystal cool below T_c, we can induce spontaneous polarization to develop along the field direction. So, we can define the c axis shown in figure 4.15.b by imposing a temporary external field. The c axis is called ferroelectric axis. The dielectric strength of the material is different for the filed applied along a-axis and c-axis. Ferroelectric crystals have large dielectric constants. Along a-axis the dielectric constant is ~4100 compared to that along c which is ~160.

Above T_c the dielectric constant usually follows a Curie-Weiss behavior which is given by

$$\varepsilon_r = \frac{C}{T - T_0},$$

where C is Curie constant and T_0 is Curie-Weiss temperature.

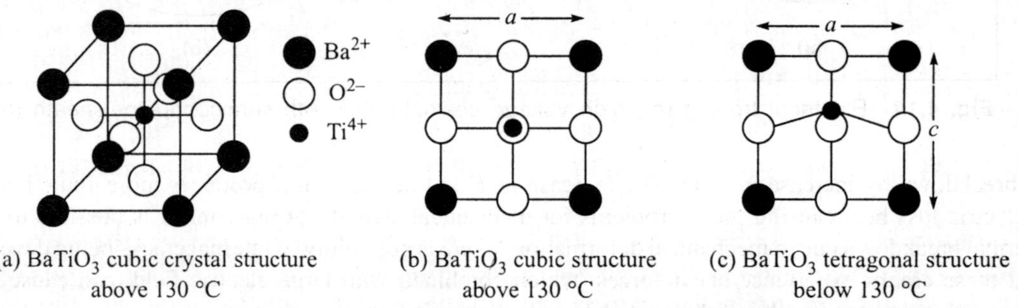

Fig. 4.15 BaTiO$_3$ at different temperatures giving different dielectric properties.

The variation of polarization with electric field is not linear for such crystals but forms a closed loop called the hysteresis loop which is similar to the ferromagnetic loop as discussed in chapter-5. The ferroelectric crystals maybe classified in to two main groups: the order-disorder and displacive group. In order-disorder group, the ferroelectric transition is associated with individual ordering of ions. These crystals contain hydrogen bond and in which the motion of protons is related to ferroelectric properties. Examples of such materials are KH_2PO_4, RbH_2PO_4 etc. The displacive group of ferroelectric materials is the one in which the ferroelectric transition is associated with displacement of a whole sub-lattice of ions of one type relative to a sub-lattice of another type. Few examples of such type of ferroelectric materials are $BaTiO_4$, $PbTiO_3$, $KNbO_3$, $LiTiO_3$, etc.

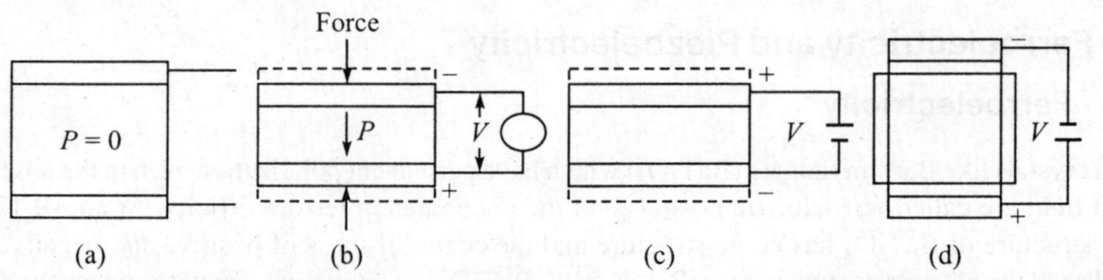

Fig. 4.16 Piezoelectric crystal with no applied field or stress (a), crystal strained by applied force (b), an applied field causes crystal to be strained (c), and direction of strain changes with change in applied field direction (d).

4.5.2 Piezoelectricity

Certain crystals become polarized under mechanical stress. Also, when an electric field is applied to these crystals, they will be mechanically deformed. These two effects are complimentary and termed as piezoelectricity.

Only certain crystals exhibit piezoelectricity because the phenomenon requires a special crystal structure–that which has no center of symmetry. The NaCl unit cell has cubic symmetry and there is no net polarization when there is no force applied and the specimen is still not polarized after it has been applied by external force as shown in figure 4.17.

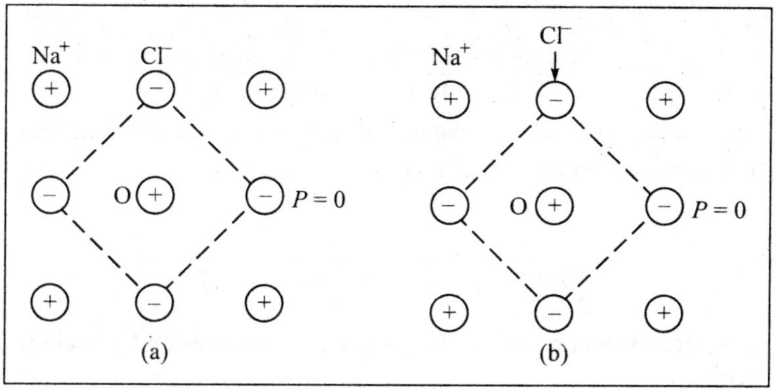

Fig. 4.17 Without external force (a) and with external force applied (b)

There are crystals that have no center of symmetry. The unit cell of such crystals is called noncentrosymmetric. In these crystals as shown in figure 4.18, in the absence of external force, the center of mass of both the positive and negative charges coincide. But when an external force is applied, the centers of masses of both the charges are shifted and there is now net polarization of the specimen. The direction of the induced polarization depends upon the direction of applied stress. If T_j is the applied mechanical stress along j direction and P_i is the induced polarization, then they are related as

$$P_i = d_{ij} T_j \tag{4.50}$$

where d_{ij} is called piezoelectric coefficient.

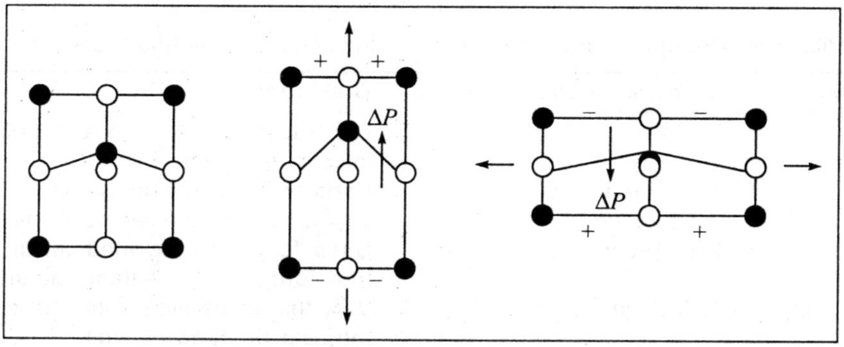

Fig. 4.18 Without applied force (a), Under force in *y*-direction *P* is along *y* (b), and under force in irection *P* is along *x* as well but in addition to *x*, *P* can be in *y* also.

The converse piezoelectric effect between an induced strain S_j along j-direction and applied electric field E_i along I-direction is given by

$$S_j = d_{ij} E_i \tag{4.51}$$

where the coefficient d_{ij} has the same meaning as before.

The electromechanical coupling between electrical and mechanical energies is denoted by coupling factor k and is defined in terms of k^2 as

$$k^2 = \frac{\text{Electrical energy converted to mechanical energy}}{\text{Electrical Input Energy}} \tag{4.52}$$

or conversely

$$k^2 = \frac{\text{Mechanical energy converted to elecrical energy}}{\text{Mechanical Input Energy}} \tag{4.53}$$

All ferroelectric materials are piezoelectric but not all the piezoelectric materials are ferroelectric.

If C and C_0 be the mechanical and electrical capacitances, then

$$k^2 = \frac{\frac{1}{2}CV^2}{\frac{1}{2}CV^2 + \frac{1}{2}C_0V^2} = \frac{C}{C + C_0} = 1 - \frac{f_r^2}{f_a^2} \tag{4.54}$$

where f_r and f_a are resonance and anti-resonance frequencies and are measured when impedance is minimum and maximum respectively.

Table 4.2 Piezoelectric materials with their *d* and *k* values

Crystal	d, mV^{-1}	k	*Application*
Quartz (crystal SiO_2)	2.3×10^{-12}	0.1	Crystal oscillators, ultrasonic transducers, filters
Barium Titanate $BaTiO_3$	190×10^{-12}	0.49	Accelerometers
PZT (lead zirconate titanate)	480×10^{-12}	0.72	Earphones, microphones, spark generators, displacement trancducers
Polyvinylidene fluoride (PVDF)	18×10^{-12}	-	Used in wide range, very cheap

4.6 Properties of Some Dielectric Materials

Table 4.3 Dielectric strength at room temperature for certain insulating material

Dielectric medium	*Dielectric strength at 60Hz*	*Comments*
Air at 1atm	31.7 kV cm^{-1}	Breakdown by electron ava-lanche by impact ionization
SF_6 gas	79.3 kV cm^{-1}	Used in high voltage circuit
Polybutene	>138 kV cm^{-1}	Liquid dielectric used as oil filler
Transformer oil	128 kV cm^{-1}	Used for insulating medium in transformer, high voltage circuit breakers
Amorphous SiO_2	10 MV cm^{-1} dc	Very thin oxide films with- out defect. Intrinsic breakdown limit
Borosilicate glass	10 MV cm^{-1} duration 10μs	Intrinsic and thermal breakdown
Polypropylene	295-314 kV cm^{-1}	Thermal breakdown or electrical treeing

Table 4.4 Experimentally determined permanent dipole moments (x 3.33 × 10^{-30}C-m)

Molecule	p_o	*Molecule*	p_o	*Molecule*	p_o
NO	0.1	HBr	0.79	H_2O	1.84
CO	0.11	NO_2	0.4	H_2S	0.93
HCl	1.04	HI	0.38	CH_3Cl	1.15

The real part of relative dielectric constant and the loss tangent of various dielectrics at a number of frequencies are presented in table 4.5 below.

Table 4.5

Material		*Frequency, Hz*				
		10^2	10^4	10^6	10^8	3×10^9
Pyranol 1467	ε'	4.42	4.0	4.4	4.08	2.84
	$10^4 \times \tan \delta$	36	4	25	1300	1200
Cab;e Oil 5314	ε'	2.25	2.25			2.22
	$10^4 \times \tan \delta$	3	0.4			18
Teflon	ε'	2.1	2.1	2.1	2.1	2.1
	$10^4 \times \tan \delta$	5	3	2	2	1.5
Polysterene	ε'	2.59	2.56	2.56	2.55	2.55
	$10^4 \times \tan \delta$	0.5	05	0.7	1.0	3.3
Polyethelene	ε'	2.25	2.25	2.25	2.25	2.25
	$10^4 \times \tan \delta$	5	3	4		3
Nylon 66	ε'	3.88	3.6	3.33	3.16	3.03
	$10^4 \times \tan \delta$	144	233	257	210	128
Bakelite BM-120	ε'	4.87	4.62	4.36	3.95	3.7
	$10^4 \times \tan \delta$	300	200	280	380	438
Glass (Cornig 0010)	ε'	6.68	6.57	6.43	6.33	6.1
	$10^4 \times \tan \delta$	77	35	16	23	360
Porcelain N0. 4462	ε'	8.99	8.95	8.95	8.95	8.9
	$10^4 \times \tan \delta$	22	6	2	4	11

Table 4.6 Static and Optical Dielectric Constants for Alkali Halides

Solid	ε_{rs}	ε_{ro}	*Solid*	ε_{rs}	ε_{ro}
LiF	9.27	1.92	KF	6.05	1.85
LiCl	11.05	2.75	KCl	4.68	2.13
LiI	11.03	3.8	KI	4.94	2.69
NaCl	5.62	2.25	NaF	6.0	1.74
RbI	5.0	2.63	RbBr	5	2.33
KBr	4.78	2.33	RbF	5.91	1.93
RbCl	5.0	2.19	LiBr	12.1	3.16
NaBr	5.99	2.62	NaI	6.6	2.91

ε_{rs} - static dielectric constant

ε_{ro} - optical dielectric constant

4.7 Solved Numerical Examples

1. The electronic polarizability of *Ar* atom is 1.7×10^{-40} Fm^2. Calculate the static dielectric constant of solid *Ar* given its density is 1.8g cm^{-3}. M_{at} = 39.95 g · mol^{-1} and $N_A = 6.02\times10^{23}$ mol^{-1}

Solution:

Number of Ar atoms per unit volume is

$$N = \frac{N_A \rho}{M_{at}} = \frac{(6.02\times10^{23}\ \text{mol}^{-1})\,(1.8\,\text{g cm}^{-3})}{39.95\,\text{g mol}} = 2.71\times10^{22}\ \text{cm}^{-3}$$

Now,
$$\varepsilon_r = 1+\frac{N\alpha_e}{\varepsilon_0} = 1+\frac{(2.71\times10^{28}\ \text{m}^{-3})\,(1.7\times10^{-40}\ \text{Fm}^2}{8.85\times10^{-12}\ \text{Fm}^{-1}}$$

If we use Clausius-Massotti equation, then

$$\varepsilon_r = \frac{1+\dfrac{2N\alpha_e}{3\varepsilon_0}}{1-\dfrac{N\alpha_e}{3\varepsilon_0}}$$ = 1.63 which is higher than the value found without using Clausius-Massotti equation. To find the correct value we should use Clausius-Massotti equation.

2. The number of electrons per unit volume of Si is 5×10^{28} m^{-3} and its dielectric constant is 11.9. Calculate:

a. Electronic polarizability due to valence electrons per Si atom (if one could portion the observed crystal polarization to individual atom);

b. If the Si crystal sample is electroded on opposite faces, by how many times the local field is greater than the applied field?

Solution:

a. Using Clausius-Massotti equation,

$$\alpha_e = \frac{3\varepsilon_0(\varepsilon_r-1)}{N(\varepsilon_r+2)} = \frac{3(8.854\times10^{-12}\ \text{Fm}^{-1})\,(11.9-1)}{5\times10^{28}\ \text{m}^{-3}(11.9+2)} = 4.17\times10^{-40}\ \text{Fm}^2$$

which is larger than electronic polarizability of an isolated Ar atom with larger number of electrons.

b. The local field is $E_{loc} = E + \dfrac{P}{3\varepsilon_0}$ And, $P = \chi_e\varepsilon_0 E = (\varepsilon_r - 1)\,\varepsilon_0 E$

Substituting for P

$$E_{loc} = E + \frac{(\varepsilon_r-1)\varepsilon_0 E}{3\varepsilon_0} = E\left(1+\frac{\varepsilon_0(\varepsilon_r-1)}{3\varepsilon_0}\right)$$

$$\frac{E_{loc}}{E} = 1+\frac{11.9-1}{3} = 4.63$$

So, the local field is 4.63 times greater than the applied field.

3. CsCl crystal has one Cs^+ and one Cl^- pair of ions per unit cell and a lattice parameter a = 0.412nm. The electronic polarizability of Cs^+ and Cl^- ions are 3.35×10^{-40} Fm^2 and 3.4×10^{-40} Fm^2 respectively. The mean ionic polarizability per ion pair is 6×10^{-40} Fm^2. Calculate the dielectric constant at low frequencies and what would its value at optical frequencies?

Solution:

Volume of the unit cell $V = a^3 = (0.412 \times 10^{-9}\,\text{m})^3$

Since one unit cell contains one ion pair, the number of ions per unit volume can be calculated as

$$N_i = \frac{1}{V} = 1.43 \times 10^{28}\,\text{m}^{-3}$$

N_i is the concentration of cation and anion separately. So, using Clausius-Massotti equation,

$$\frac{\varepsilon_r - 1}{\varepsilon_r + 2} = \frac{1}{3\varepsilon_0}\{N_i\alpha_e(\text{Cs}^+) + N_i\alpha_e(\text{Cl}^-) + N_i\alpha_i\}$$

$$\frac{\varepsilon_r - 1}{\varepsilon_r + 2} = \frac{(1.43 \times 10^{28}\,\text{m}^{-3})(3.35 \times 10^{-40} + 3.4 \times 10^{-40} + 6 \times 10^{-40})\text{Fm}^2}{3 \times 8.854 \times 10^{-12}\,\text{Fm}^{-1}}$$

And solving for dielectric constant we get $\varepsilon_r = 7.56$

At optical frequencies, the ionic polarization is too sluggish to allow polarization to contribute to ε_r. Thus the dielectric constant will be determined only by electronic polarization and given by

$$\frac{\varepsilon_{ro} - 1}{\varepsilon_{ro} + 2} = \frac{1}{3\varepsilon_0}\{N_i\alpha_e(\text{Cs}^+) + N_i\alpha_e(\text{Cl}^-)\}$$

$$\frac{\varepsilon_{ro} - 1}{\varepsilon_{ro} + 2} = \frac{(1.43 \times 10^{28}\,\text{m}^{-3})(3.35 \times 10^{-40} + 3.4 \times 10^{-40})\text{Fm}^2}{3 \times 8.854 \times 10^{-12}\,\text{Fm}^{-1}}$$

And solving for ε_{ro}, we will get $\varepsilon_{ro} = 2.71$.

The experimental values of dielectric constants at low and optical frequencies are 7.2 and 2.62 respectively.

4. Calculate the heat generated per second due to dielectric loss per cubic centimeter of XLPE and alumina Al_2O_3 at 60 HZ and 1MHz at a field of 100kV cm^{-1}. ε_r' and tan δ for XLPE are 2.3 and 3×10^{-4} and for alumina – 8.5 and 10^{-3} respectively at 60 Hz. Similarly at 1 MHz, these values are 2.3 and 4×10^{-4} and 8.5 and 10^{-3} respectively.

Solution:

Power dissipated per unit volume is given by

$$P_{vol} = \omega E^2 \varepsilon_0 \varepsilon_r' \tan \delta$$

For XLPE at 60Hz, $P_{vol} = 230.3\,\text{Wm}^{-3}$ And, similarly for alumina at 60Hz,

$$P_{vol} = 5\,\text{Wm}^{-3}$$

At 1 MHz the corresponding power losses per unit volume are 2840 Wm^{-3} for XLPE and 330 Wm^{-3} for alumina.

5. The optical index of refraction and the dielectric constant for glass are 1.45 and 6.5 respectively. Calculate the percentage of ionic polarizability.

Solution:

We have from Clausius-Massotti equation

$$\frac{\varepsilon_r - 1}{\varepsilon_r + 2} = \frac{N(\alpha_e + \alpha_i)}{3\varepsilon_0}.$$

At optical frequencies, we have

$$\frac{n^2-1}{n^2+2}=\frac{N\alpha_e}{3\varepsilon_0}$$

Now, dividing the second relation by the first, we have

$$\frac{n^2-1}{n^2+2}\times\frac{\varepsilon_r+2}{\varepsilon_r-1}=\frac{\alpha_e}{\alpha_0+\alpha_i}$$

So, percentage of ionic polarizability is

$$\frac{\alpha_i}{\alpha_e+\alpha_i}\times 100=\left[1-\frac{\alpha_e}{\alpha_e+\alpha_i}\right]\times 100$$

$$=\left[1-\left(\frac{1.1025}{4.1025}\right)\left(\frac{8.5}{5.5}\right)\right]\times 100=57.98\%$$

Exercise

1. A glass dielectric has dielectric constant 2.6 and loss tangent 7×10^{-5} at 1MHz. Calculate the loss of power per unit volume if the signal applied to the glass has peak amplitude of 0.71V/m. [5.05×10^{-9} W/m^3]
2. Calculate dielectric constant of a lead ceramic zirconate capacitor at 215°C if its capacitance is 2×10^{-9}F in pellet form, with radius 0.8cm and thickness 1.5mm. [1686.82]
3. Prove that heat loss in a dielectric is half its maximum value at which imaginary part of the relative permittivity is maximum using single relaxation time of dipoles.
4. Calculate the dielectric constant of Barium Strontium Titanate at 100°C if its Curie-Weiss temperature is 71°C and Curie constant 1.3×10^5°C. [4482.76]
5. Determine the percentage of ionic polarizability of water for which refractive index and dielectric constant are 1.33 and 8.1 respectively. [71%]
6. Determine real and imaginary parts of dielectric constant of a dielectric sample at 1kHz if the relaxation time for the material is 10^{-6}s. Given that the dielectric constant at low frequency and high frequency are 8 and 2 respectively. [7.99, 0.038]

Magnetic Materials

5
Chapter

5.1 Ferromagnetism, Ferrimagnetism and Paramagnetism

Many electrical engineering devices such as inductors, transformers, rotating machines, etc are based on the utilization of magnetic properties of materials. To understand magnetic materials and magnetism properly we need to know the following quantities first.

5.1.1 Magnetic Dipole Moment

A current carrying loop with A as enclosed area has a current I circulating through it as shown in figure 5.1. u_n is unit vector coming out from the area A. The direction of u_n is such that looking about it, the current circulates clockwise. Thus the magnetic dipole moment is defined by

$$\mu_m = IA\,u_n \tag{5.1}$$

In case of a coil having N turns, the magnitude of dipole moment

$$\mu_m = \mathrm{N\,IA} \tag{5.2}$$

When a magnetic dipole is placed in a magnetic field, it experiences a torque that tries to rotate the magnetic moment to align its axis with the magnetic field. A magnetic dipole moment creates magnetic field just like a bar magnet.

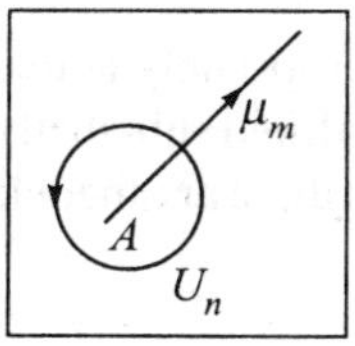

Fig. 5.1 A current loop

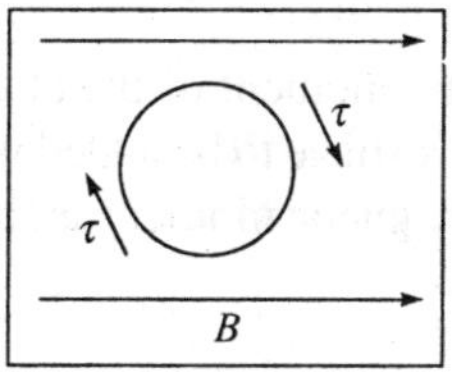

Fig. 5.2 A dipole moment in a magnetic field

5.1.2 Atomic Magnetic Moments

An orbiting electron in an atom behaves much like a current loop and has a magnetic dipole moment associated with it called orbital magnetic moment μ_{orb} as shown in figure 5.3. If ω is the angular frequency of the electron, then the current I due to the orbiting electron is

$$I = \text{Charge flowing per unit time} = \frac{-e}{\text{period}} = \frac{-e\omega}{2\pi} \tag{5.3}$$

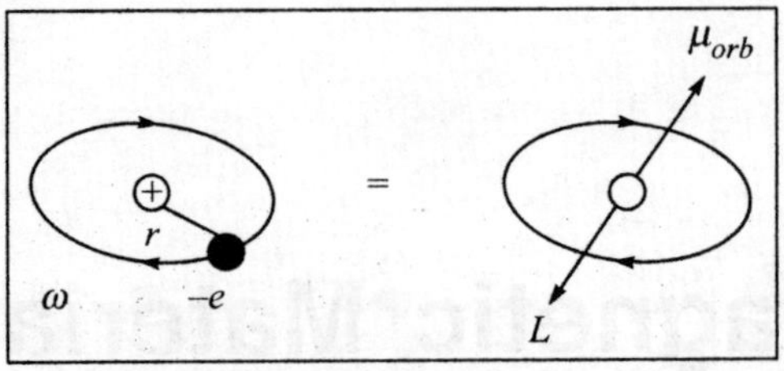

Fig. 5.3 An orbiting electron is equivalent to a magnetic dipole moment m_{orb}

If r is the radius of the orbit, then the magnetic dipole moment is

$$\mu_{orb} = I(\pi r^2) = \frac{-e\omega r^2}{2} \tag{5.4}$$

Angular momentum of the electron is

$$L = (m_e v)r = m_e(\omega r)r = m_e \omega r^2 \tag{5.5}$$

which is quantized and is represented as

$$L = l\hbar$$

where l is the orbital quantum number.

So orbital magnetic moment is

$$\mu_{orb} = \frac{-e}{2m_e} L \tag{5.6}$$

where $e/2m_e$ is called gyromagnetic ratio, which is the ratio of magnetic moment to orbital magnetic moment

and $\frac{e\hbar}{2m_e}$ is called Bohr magneton which is equal to 9.273×10^{-24} $A.m^2$ or $J.T^{-1}$. The negative sign indicates that orbital magnetic moment is in opposite direction to angular momentum.

The intrinsic angular momentum of electron denoted by S gives rise to spin magnetic moment μ_{spin}.

$$\mu_{spin} = -\frac{e}{m_e} S = -s\hbar \frac{e}{2m_e} \tag{5.7}$$

The overall magnetic moment of the electron consists of μ_{orb} and μ_{spin} appropriately added. They are both vector quantities, so have to be added vectorically. Only the electrons in unfilled sub-shells contribute to the overall atomic magnetic moment as for each electron in closed sub-shells there another electron with opposite spin.

5.1.3 Magnetization Vector M

The magnetic field inside the solenoid with free space inside it B_0 is given by

$$B_0 = \mu_0 n I = \mu_0 I' \tag{5.8}$$

where n = number of turns per unit length,

I = current through the solenoid, A

μ_0 = absolute permeability of free space = $4\pi \times 10^{-7}$ H/m

I' = current per unit length = nI

If a cylindrical material is placed inside the solenoid, the magnetic field will be changed from B_0 to B. Each atom of the material responds to the applied field B_0 and develops a net magnetic moment μ_m along the applied field. The medium is thus magnetized. Magnetic vector M describes the extent of magnetization of the medium. So, magnetization vector M is defined as magnetic dipole moment per unit volume. If there are N atoms in a small volume of V and each atom, i, has a magnetic moment μ_{mi}, then M is defined as

$$M = \frac{1}{V}\sum_{i=1}^{N}\mu_{mi} = n_{atomic}\mu_{av} \tag{5.9}$$

where n_{atomic} = number of atoms per unit volume

μ_{av} = average magnetic dipole moment per atom

It can be assumed that each atom acquires a magnetic moment μ_{av} along $\boldsymbol{B_0}$ which can be viewed as elementary current loops at the atomic scale. The elementary current loops are due to electronic currents within the atom and arise both from the orbital and spin motion. The current in the bulk cancel out each other, so there is no net current within the bulk. Only the surface current will be there. The surface currents are induced by the magnetization of the medium by the applied magnetic field and depend on the magnetization vector M of the specimen. Total magnetic moment of the cylindrical specimen is

$$\text{Total Magnetic Moment} = MV = MAL \tag{5.10}$$

If I_m is the magnetization current on the surface per unit length, then the total circulating current is $\boldsymbol{I_m L}$ and total magnetic moment is by definition

$$\text{Total Magnetic Moment} = \text{Total Current x Cross Sectional area} = I_m AL$$

$$MAL = I_m AL$$

$$M = I_m \tag{5.11}$$

Magnetizing current $\boldsymbol{I_m}$ is not due to the flow of free charge carriers as in a current carrying conductor. This is due to the localized electronic currents within the atoms of the solid at the surface. For a cylindrical specimen, $\boldsymbol{I_m}$ is perpendicular to $\boldsymbol{M}$.

5.1.4 Magnetizing Field or Magnetic Field Intensity

When a medium is inserted within the solenoid as shown in figure 5.4, the magnetic field inside the solenoid changes from $\boldsymbol{B_0}$ to $\boldsymbol{B}$, and $\boldsymbol{B}$ is given by

$$B = \mu_0 (I' + I_m) = B_0 + \mu_0 I_m = B_0 + \mu_0 M \tag{5.12}$$

where $\boldsymbol{B_0}$, $\boldsymbol{B}$ and $\boldsymbol{M}$ all are vector quantities.

The field inside the magnetized material is the sum of the applied field $\boldsymbol{B_0}$ and a contribution from the magnetization $\boldsymbol{M}$ at the material. The magnetizing field is defined as

$$H = \frac{1}{\mu_0} B - M$$

$$H = \frac{1}{\mu_0} B_0 \tag{5.13}$$

H is simply related to the external conduction current of free charge carriers.

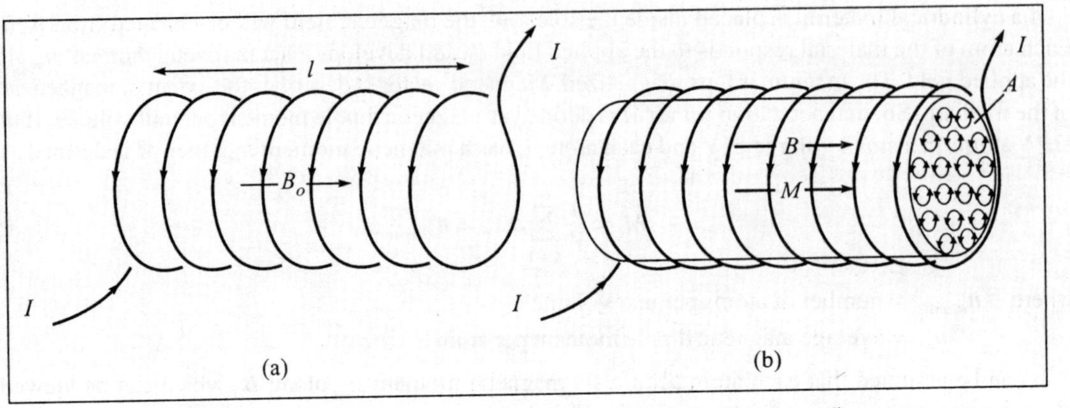

Fig. 5.4 Magnetic field of a solenoid with and without medium.

For a solenoid

$$B_0 = n\mu_0 I$$
$$H = nI \tag{5.14}$$

Magnetizing field ***H*** is the cause the cause and magnetic field ***B*** is its effect. ***H*** solely depends upon external conduction currents only whereas the magnetic field ***B*** depends on the magnetization of the material.

5.1.5 Magnetic Permeability and Susceptibility

The magnetic permeability of the medium at any point is defined as the magnetic field per unit magnetizing field

$$\mu = \frac{B}{H} \tag{5.15}$$

μ represents to what extent a medium is permeable by a magnetic field.

Relative permeability of the medium is defined as the fractional increase in the field with respect to the field in free space when a material medium is introduced.

$$\mu_r = \frac{B}{B_0} = \frac{B}{\mu_0 H} \tag{5.16}$$

Relating expression (5.16) with (5.15) we will get

$$\mu = \mu_0 \mu_r \tag{5.17}$$

where μ = absolute permeability of the medium

μ_r = relative permeability of the medium

μ_0 = absolute permeability of the free space = $4\pi \times 10^{-7}$ H.m^{-1}

The magnetization ***M*** produced in the material depends on the net magnetic field ***B***. It should be natural to relate magnetization ***M*** to magnetic field ***B***. But, for historic reasons, magnetization ***M*** is related to magnetizing field ***H***. ***M*** is related to ***H*** as

$$M = \chi_m H \tag{5.18}$$

where χ_m = Magnetic susceptibility

The relationship given in (5.18) is obeyed by all materials.

Rewriting equation (5.12)

$$B = B_0 + \mu_0 M$$
$$B = \mu_0 H + \mu_0 M$$
$$B = \mu_0 H + \mu_0 \chi_m H$$
$$B = (1 + \chi_m)\, \mu_0 H$$
$$B = \mu_0 \mu_r H \quad (5.19)$$

where $\mu_r = 1 + \chi_m$ (5.20)

Expression (5.20) relates the relative permeability with magnetic susceptibility of the material.

5.1.6 Ferromagnetism

Ferromagnetic materials can possess large permanent magnetization even in the absence of an applied field. The magnetic susceptibility χ_m is positive and very large and depends on the applied field intensity. The relationship between magnetization $\boldsymbol{M}$ and $\mu_0 \boldsymbol{H}$ is highly non-linear. At sufficiently high fields, magnetization $\boldsymbol{M}$ saturates. The origin of the ferromagnetism is the quantum mechanical exchange interaction between the constituent atoms that results in regions of material possessing permanent magnetization. The ferromagnetic crystal has magnetic moments of all crystal aligned in an orderly manner so as to give rise to the net magnetization vector $\boldsymbol{M}$. Ferromagnetism occurs below a critical temperature T_c called Curie temperature. At temperatures above this ferromagnetism is lost, i.e., paramagnetism is obtained obeying Curie-Weiss Law $\chi = \dfrac{C}{T - T_0}$. We will discuss about ferromagnetism in detail later on.

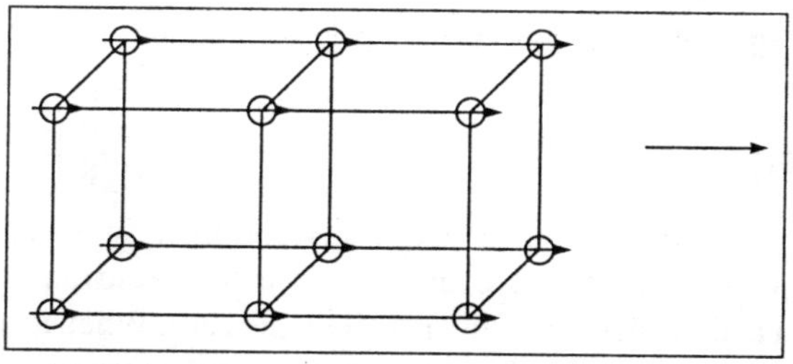

Fig. 5.5 Magnetized region of Ferromagnetic material

5.1.7 Antiferromagnetism

Antiferromagnetic materials have small positive susceptibility. In the absence of external field, there is no net magnetization of the material. They possess magnetic ordering in which the magnetic moments on alternating atoms in the crystals align in opposite directions. The opposite alignments of magnetic moments are due to quantum mechanical exchange interaction. Antiferromagnetism occurs below a critical temperature called Néel temperature. Above this they are paramagnetic.

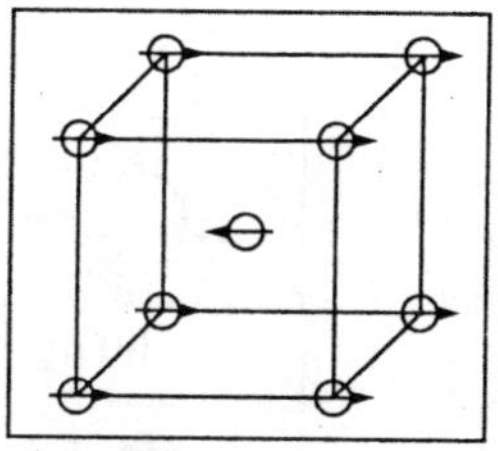

Fig. 5.6 Antiferromagnetic crystal (Cr)

Antiferromagnetic substances are feebly magnetized when subjected to a strong magnetic field. Examples of Antiferromagnetic substances are MnO, FeO, CoO, NiO, $FeCl_3MnO_4$, Cr, Mn, MnS, etc.

5.1.8 Ferrimagnetism

Ferrimagnetic materials exhibit magnetic behavior similar to ferromagnetism below a critical temperature called Curie temperature T_C. As shown in figure 5.7 above all *A* atoms has their spins aligned in one direction and all *B* atoms has their spins aligned in opposite direction. Magnetic moment of atom *A* is greater than that of atom *B*. Due to this there will be net magnetization *M* in the crystal even in the absence of external applied filed. These materials are non-conducting, so they do not suffer from eddy current losses. This is the reason for these materials being used for high frequency electronic application. Ferrimagnetic materials have larger net magnetization as compared to antiferrimagnetic materials. These materials are generally known as ferrites and represented by XFe_3O_4 where X = Mg, Cu, Mn, Ni, Zn, Cd, etc.

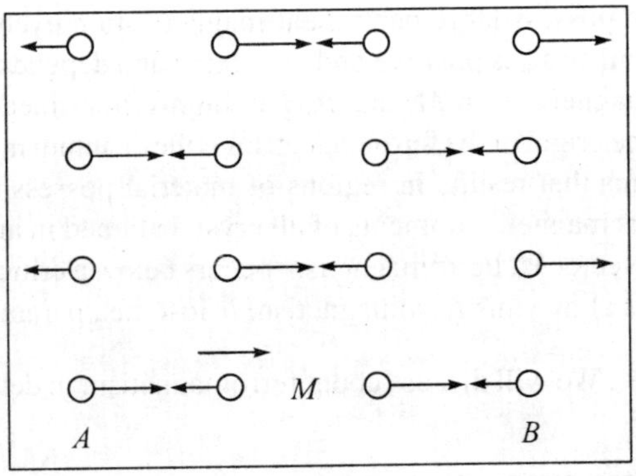

Fig. 5.7 Magnetic ordering in ferrimagnetic crystal

5.1.9 Paramagnetism

Paramagnetic materials have small positive magnetic susceptibility. In the absence of external magnetic field, molecular moments (atomic moments) are randomly oriented due to random collision of molecules. So the average dipole moment and the net magnetization both are zero. When an external field is applied, μ_{av} non-zero and depends upon the applied field $\mu_0 H$ and hence the magnetization is also non-zero and is equal to $\chi_m \boldsymbol{H}$. Magnetization $\boldsymbol{M}$ increases with $\mu_0 \boldsymbol{H}$ but decreases with increase in temperature. At higher

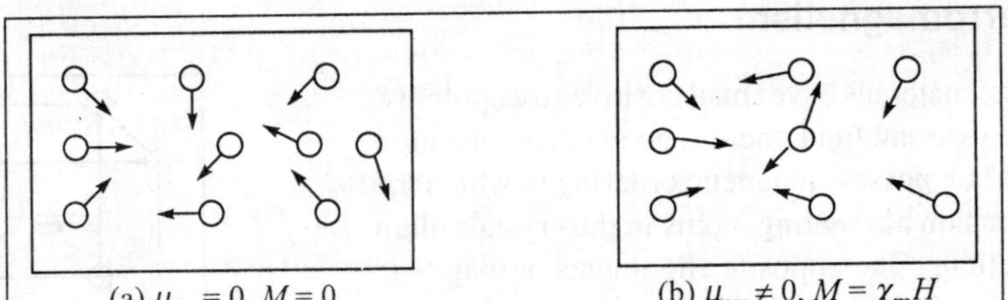

(a) $\mu_{av} = 0, M = 0$ (b) $\mu_{av} \neq 0, M = \chi_m H$

Fig. 5.8 Without external field (a), with external field (b) temperatures, the molecular collisions destroy the alignment of molecular magnetic moments with applied field, i.e., Curie law ($\chi_m = C/T$) is obeyed. When a paramagnetic substance is placed in a non-uniform magnetic field, the induced magnetization ***M*** is along the direction of ***B*** and there is a net force toward greater fields. Many gases and metals are paramagnetic in nature. The typical value of magnetic susceptibility is $\chi_m = 1.2 \times 10^{-5}$ for metals and $\chi_m = 2.1 \times 10^{-6}$ for oxygen. Other examples are Pt, Al, Cr, Mn and dilute solutions of ferromagnetic materials.

5.1.10 Diamagnetism

Materials with negative magnetic susceptibility are termed as diamagnetic substances. Their relative permeability is slightly less than unity. When diamagnetic materials are placed in magnetic field, the magnetization vector ***M*** is in opposite direction to the applied field. This causes the magnetic field within the material to be less than the applied field. The negative susceptibility can be interpreted as the diamagnetic substance trying to expel the applied field from within the material. Covalent crystals and many ionic crystals are diamagnetic because the constituent atoms have no unfilled sub-shell. Superconductors are perfect diamagnet with $\chi_m = -1$ and totally expel the applied field. Other examples of diamagnetic substances are Bi, Sb, Au, Cu, Hg, water, air, alkalihalides, organic materials, many polymers, covalent solids like Si, Ge, diamond, etc. The average value of magnetic susceptibility of diamagnetic materials is in the range of -10^{-6}.

5.2 Domain Structure

5.2.1 Introduction

A single piece of iron does not necessarily possess a net permanent magnetization in the absence of an applied filed. The reason for the absence of net magnetization is due to formation of magnetic domains that cancel each other. A magnetic domain is a region of the crystal in which all the spin magnetic moments are aligned to produce a magnet in one direction only.

In figure 5.9(a), a single crystal iron has a permanent magnetization as a result of aligning of all atomic spins. The potential energy called magnetostatic energy of the magnet can be reduced by dividing the crystal into two regions (domains) with opposite magnetization. The external magnetic field lines are reduced and there is now less potential energy stored in the magnetic field. There are only field lines at the end. The boundary between two domains is called domain wall or Bloch wall where the magnetization changes from one direction to the opposite. The domain wall is not simply one atomic spacing but has a finite thickness, which for iron is of the order of 0.1 micron, i.e., several hundred atom spacing.

The magnetostatic energy associated with the field lines at the ends (figure 5.9.b) can be further reduced by closing the ends with sideway domains with magnetization at 90°. These sideway domains at 90° with the main domains are called end domains. Although the magnetostatic energy has decreased, the potential energy in the walls has increased due to the addition of walls. The creation of magnetic domain continues spontaneously until the potential energy reduction in creating an additional domain is same as the increase in addition wall. The specimen then possesses minimum potential energy and is in equilibrium with no net magnetization. The sizes, shapes and distribution of domains depend on a number of factors including the size and shape of the whole specimen. For iron particles of dimension less than 0.01micron, the increase in potential energy in creating a domain wall is too costly and these particles are single domains and hence always magnetized.

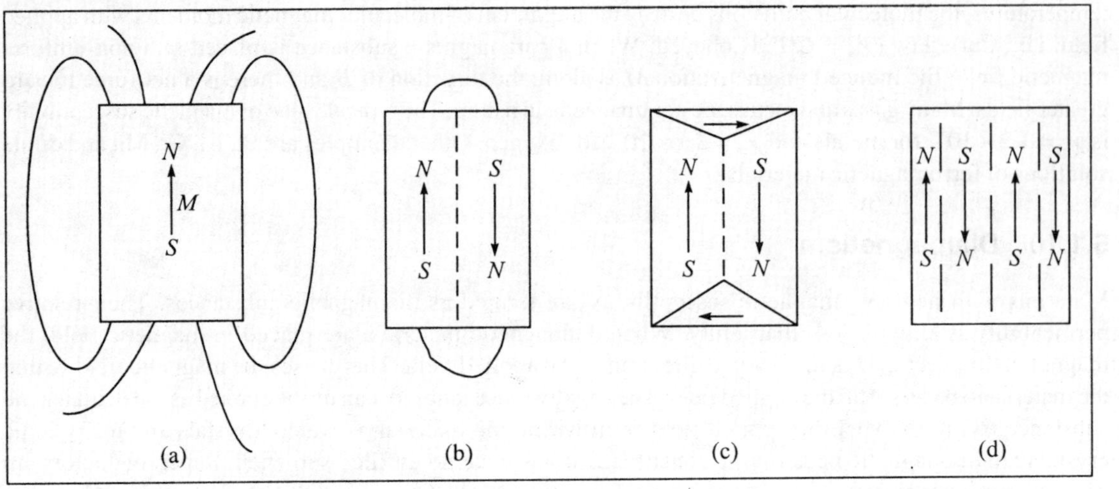

Fig. 5.9 Domain structure in ferromagnetic material. Magnetized bar of ferromagnet (a). Formation of two domains with opposite magnetization (b). Domains of closure fitting eliminating the external field lines(c). Specimen with several domains and closure domains (d).

The magnetization of the domains is normally along the one of the preferred directions in which the atomic spin alignment is easiest. For iron, the magnetization is easiest along one of the six <100> directions (cube edges). The domains have magnetizations along these directions called easy directions. The magnetization of the crystal along an applied field occurs by growth of the domains with magnetizations along the applied field.

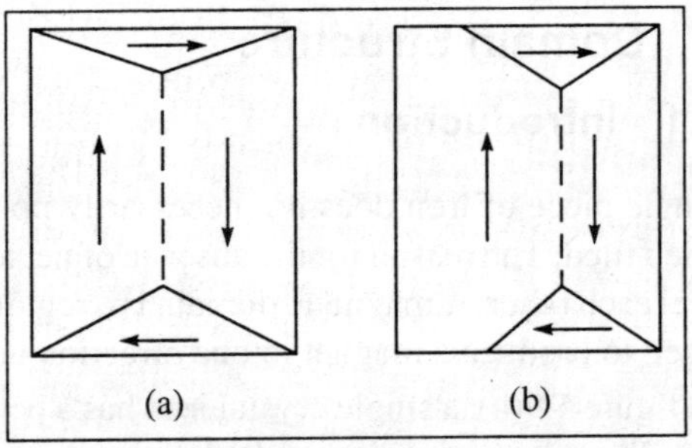

Fig. 5.10 Domain structure in the absence of applied field (a), in the presence of field (b)

For simplicity magnetization is taken along an easy direction. The Bloch wall between domains *A* and *B* migrates toward right enlarging domain *A* and shrinking domain *B*. The crystal has net magnetization ***M*** along applied field ***H***. The migration of Bloch wall is caused by the spins in the wall and also spins in the region *B* adjacent to the wall being gradually rotated by the field. The magnetization process involves the motions of Bloch walls in the crystal.

5.2.2 Domain Walls

The spin magnetic moments rotate across a domain wall which is several hundreds atom spacing thick. It is seen that the neighboring spin moments in the wall rotate gradually, and over several hundred atomic spacing the magnetic moments reach a rotation of 180°. Exchange forces between neighboring atomic spins favor very little relative rotation. Had it been left to the exchange forces alone, relative rotation of atomic spins would be so minute that the wall would have to be very thick (infinitely thick) to achieve a rotation of 180°. However, magnetic moments that are oriented away from the easy direction possess excess energy called the anisotropy energy. If the wall is thick, then it will contain many magnetic moments

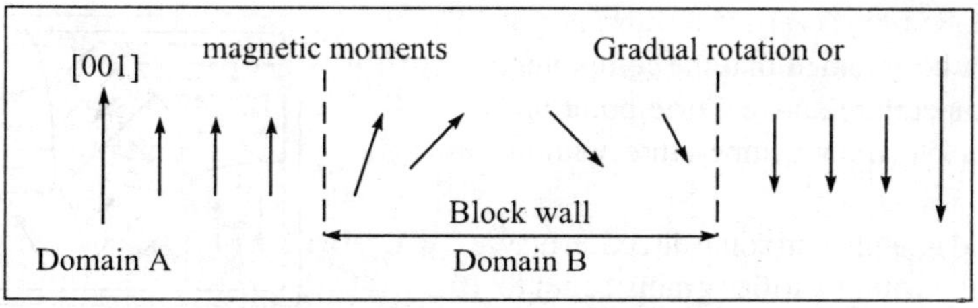

Fig. 5.11 Domains and domain wall

rotated away from the easy directions and there would be substantial anisotropy energy in the wall. Minimum anisotropy energy in the wall is obtained when the magnetic moment changes direction by 180° from the easy direction along +z to another easy direction –z without any intermediate rotations away from *z*. This requires a wall of one atomic spacing. In reality, the domain wall thickness is a compromise between the exchange energy requiring thick wall and anisotropy energy requiring a thin wall> the equilibrium wall thickness is that which minimizes the total potential energy which is the sum of exchange energy and anisotropy energy within the wall. For iron the thickness of the domain wall is of the order of 0.1micron and for cobalt it is smaller due to higher anisotropy energy.

5.2.3 Magnetostriction

We can create strain in a ferromagnetic crystal by applying suitable stress along a certain direction thereby changing the interatomic spacing not only along this direction but also in other directions causing the change in exchange interaction between atomic spins. This will lead to change in magnetization properties of crystal. In the converse effect, magnetization of the crystal generates strains or changes in the physical dimensions of the crystal. When an iron crystal is magnetized along [111] direction by a strong field (this being hard direction only strong field can magnetize along this), the atomic spins within the domains are rotated from their easy directions toward the hard direction. These electron spin rotations involve changes in the electron charge distributions around the atoms and therefore affect the interatomic bonding and hence the interatomic spacing. The longitudinal strain $\Delta L/L$ along the direction of magnetization is called the magnetostrictive constant λ. This constant depends on the crystal direction and may be positive (extension) or negative (contraction). The crystal lattice strain energy associated with magnetostriction is called the magnetostrictive energy, which is typically less than the anisotropy energy.

Magnetostriction is responsible for humming noise in transformers. As the core of a transformer is magnetized in one way and then in the opposite by an alternating current, the alternating changes in longitudinal strain vibrate the surrounding environment, air, oil, and so forth, and generate an acoustic noise at twice the main frequency and its harmonics. The magnetostriction constant can be controlled by alloying metals. The saturation strain for Ni along an easy direction is negative whereas that of iron is positive. So, by alloying 85% Ni with 15% Fe the resulting magnetostrictive constant is zero.

5.3 Hysteresis loop, Eddy Current Losses

The majority of magnetic materials used are polycrystalline having microstructure consisting of many grains of various sizes and orientations depending upon the preparation and thermal history of the specimen.

In unmagnetized polycrystalline sample, each crystal grain will possess domains, which will depend on the shape and size of the grain and to some extent, on the magnetization of the neighboring grains. The net magnetization of the sample is zero in the absence of external field. It can be assumed that the component was heated to a temperature above Curie point and then allowed to cool to room temperature without applied field.

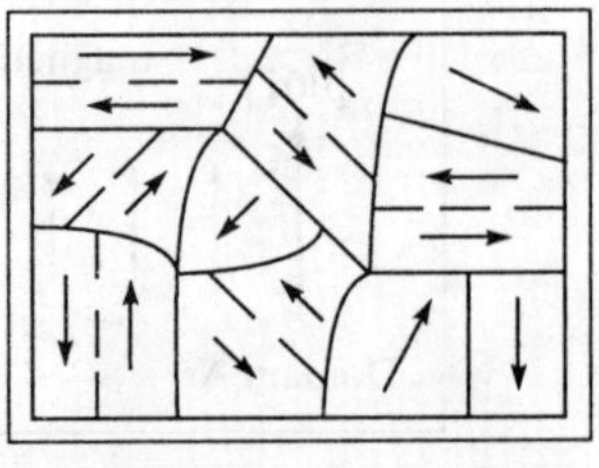

Fig. 5.12 Domains in the grains of an unmagnetized polycrystalline sample

When a small field is applied in some direction (say +x), the domain walls within various grains begin to move small distances, and domain with net magnetization ***M*** along +x grows a little larger at the expense of those pointing away from the field.

At the start there is very small net magnetization along ***H*** as indicated by region *o-a*. As we increase the field, the domain motions extend to larger distances and the walls encounter various obstacles such as crystal imperfections, impurities and so on, which try to attract the wall and thereby hinder the motion. A domain wall that is stuck or pinned at an imperfection at a given field cannot move until the field increases sufficiently to provide the necessary force to snap the wall free, which then suddenly surges to the next obstacle. Sudden snapping of the wall causes lattice distortion creates lattice waves that carry off some of the energy. Sudden changes in magnetization induçe eddy currents that dissipate energy via joule heating (I^2R). The domain walls have finite electrical resistance. These processes involve energy conversion to heat and are irreversible. Sudden jerks in wall motion leads to a small jump in magnetization of the specimen as the magnetizing field is increased. This phenomenon is termed as ***Barkhausen effect***.

As we continue to increase the field, magnetization vector ***M*** continues to increase by jerky domain wall motions that enlarge domain with favorably oriented magnetizations and shrink away those with magnetization pointing away from the applied field. Eventually domain wall motions leave each crystal grain with a single domain and magnetization in one of the easy directions.

If we decrease and remove the field, the magnetization in each grain will rotate to align parallel with the nearest easy direction in that grain. In some grain additional small domains may develop that reduce the magnetization within the grain (point *e*). This process from point *d* to *e* whereby the magnetizing field is reduced to zero leaves the specimen with the net permanent magnetization called ***remanent*** or ***residual magnetization*** M_r. If now a magnetizing field is applied in the reverse direction, the magnetization of the specimen would decrease and eventually at a sufficiently large applied field, magnetization ***M*** becomes zero. This means the specimen has been totally demagnetized. The magnetizing field required to totally demagnetize the sample is called ***coercive field*** and is denoted by H_c. Coercive field represents the opposition of the sample to demagnetization. The demagnetization process involves the generation of new domains, where nucleation at various crystal imperfections cancels out overall magnetization.

If the magnetization is further increased in reverse direction, the process from point *e* will be similar to that when magnetizing the sample. Point *g* will represent saturation magnetization $-M_{sat}$ in reverse direction. Decreasing the magnetizing field from $-H_{sat}$ to zero again leaves residual magnetization in the specimen and to eliminate this we need to apply magnetizing field in reverse direction (now in positive *x* direction). The complete ***M*** versus ***H*** curve obtained by changing magnetizing field from positive *x* to negative *x* for corresponding values of magnetization is a closed loop and called ***hysteresis loop***. In both the positive and negative directions there are saturation magnetization M_{sat} and saturation magnetizing field H_{sat}. The initial magnetization curve for a previously unmagnetized specimen ***o-a-b-c-d*** is called initial magnetization curve.

Under very small fields (a); boundary motions occur in sudden jerks (b); nearly all grains are single domains (c); magnetization in individual grains has to be rotated to align with ***H*** (d); after the external field

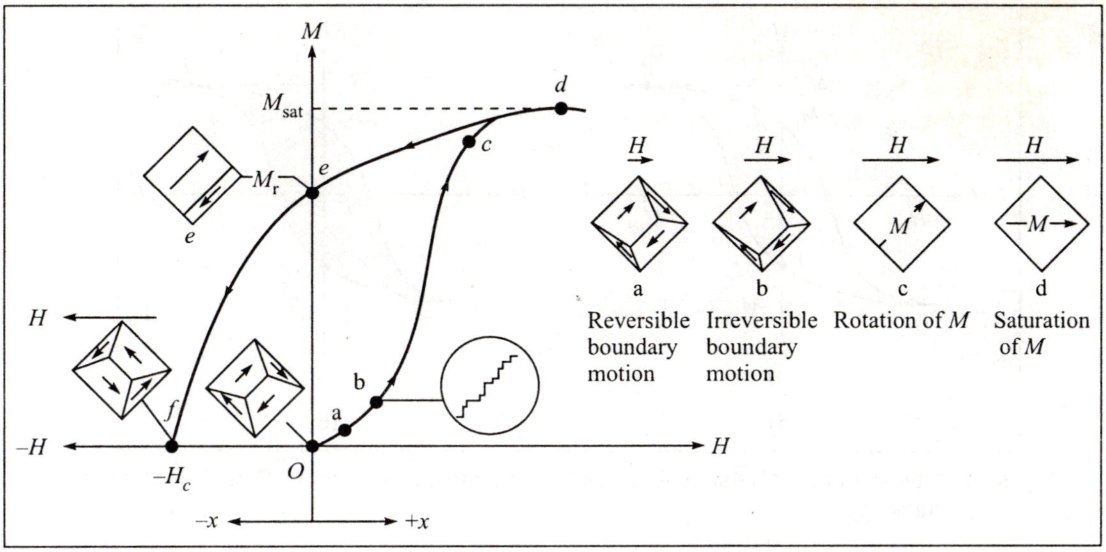

Fig. 5.13 ***M*** vs. ***H*** behavior of a previously unmagnetized polycrystalline iron specimen

has been removed (e); the remaining magnetization $\boldsymbol{M_r}$ has been removed by applying magnetizing field $\boldsymbol{H_c}$ in reverse direction (f).

We know the following relationship $B = \mu_0 M + \mu_0 H$ which leads to hysteresis loop in the magnetic field versus magnetizing field behavior. The very slight increase in $\boldsymbol{B}$ with $\boldsymbol{H}$ when $\boldsymbol{M}$ is in saturation is due to the permeability of free space. The area enclosed within $\boldsymbol{B}$ versus $\boldsymbol{H}$ hysteresis loop shown by hatched line in figure 5.14 (b) is the measure of energy dissipated per unit volume per cycle of applied field variation. If the magnetic field in the material does not reach $\boldsymbol{B_{sat}}$ but instead reaches some maximum value $\boldsymbol{B_m}$ for a magnetizing field $\boldsymbol{H_m}$, still there will be a hysteresis loop formed since the magnetization and demagnetization are nonreversible and do not trace each other. Only the area of the loop will be smaller than that for $\boldsymbol{B_{sat}}(\boldsymbol{M_{sat}})$ versus $\boldsymbol{H_{sat}}$. So the shape of hysteresis loop depends in addition to shape and size of the specimen on the magnitude of the applied field.

The energy dissipated during magnetization and demagnetization can be written as

$$dE_{vol} = HdB$$

When the magnetic field changes from B_1 to B_2

$$E_{vol} = \int_{B_1}^{B_2} HdB \tag{5.21}$$

The hysteresis power loss for steel cores can be empirically expressed as

$$P_h = KfB_{rn}{}^n \tag{5.22}$$

where P_h = hysteresis power loss, W

f = frequency of applied voltage, Hz

B_m = maximum magnetic field (flux density), T

n = 1.6

K = constant which depends upon core material (typically K = 150.7)

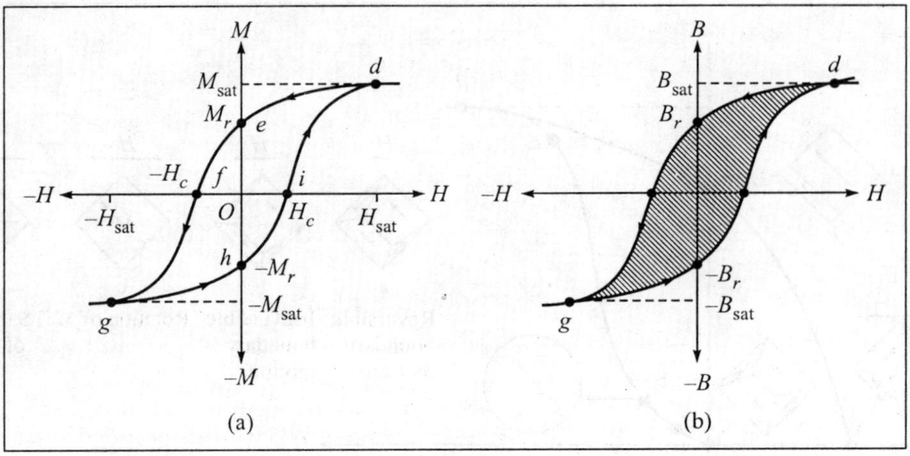

Fig. 5.14 A typical ***M*** vs. ***H*** hysteresis curve (a), and the corresponding ***B*** vs. ***H*** hysteresis curve (b).

5.4 Soft Magnetic Materials

When a magnetic material is magnetized and then demagnetized, energy loss takes place that is given by the area of the hysteresis loop. The size of the loop is different for different materials. There are materials, which can be easily magnetized and then demagnetized. And, other materials are hard both to magnetize and demagnetize. So, according to the *B-H* behavior, magnetic materials are classified into soft magnetic materials and hard magnetic materials.

Soft magnetic materials are easy to magnetize and demagnetize, and hence require relatively low magnetic field intensities. In other words, their *B-H* loops have small area, so the power loss per unit cycle is small. They are most suited for applications where repeated cycles of magnetization and demagnetization are involved. Electric motors, transformers, inductors have magnetic fields varying cyclically. So they require magnetic materials with low loss (small hysteresis loop). Electromagnetic relays which have to be switched on and off frequently for their assigned operation also require the iron core to be magnetized and demagnetized correspondingly. So, they require soft magnetic materials as well.

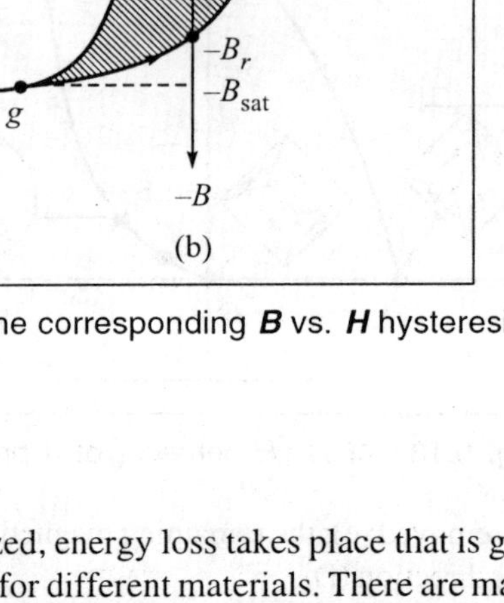

Fig. 5.15 ***B-H*** curve for typical soft and hard magnetic materials

An ideal soft magnetic material should have zero coercivity (coercive field), a very large saturation magnetization M_{sat} (correspondingly B_{sat}), zero remanent magnetization M_r (and of course B_r), zero hysteresis loss and high initial and relative permeability μ_{ri} and maximum relative permeability μ_{rmax}.

Table 5.1 *Selected soft magnetic materials (E_h is hysteresis loss,i.e., energy dissipated per unit volume per cycle, J/m^3 at B_m = 1 T)*

Magnetic Material	$\mu_0 H_c$, T	B_{sat}, T	B_r, T	μ_{ri}	μ_{rmax}	E_h, Jm^{-3}	Applications
Ideal soft iron	0	Large	0	Large	Large	0	Transformer cores, inductors, magnetic recording heads, relays, electromagnet cores, machines
Iron (commercial grade, 0.2% impuriries	$<10^{-4}$	2.2	<0.1	150	10^4	250	Due to large eddy current losses used only in some elect-magnets and relays
Silicon Iron (2-4% Si)	$<10^{-4}$	2	0.5-1	1000	10000 - 400000	30 - 100	Wide range of ac machinery
Supermalloy (79%Ni, 15%Fe, 5%Mo, 0.5%Mn)	2×10^{-7}	0.7–0.8	<0.1	10^5	10^6	<0.5	High permeability low loss electrical devices, e.g. speciality transformers, magnetic amplifier
Permalloy (78.5%Ni, 21.5%Fe)	5×10^{-6}	0.86	<0.1	8000	10^5	<0.1	Low loss electrical devices, audio transformers, HF transformers, recording heads, filters
Glassy metals (Fe – Si – B)	2×10^{-6}	1.6	$<10^{-6}$	–	10^5	20	Low loss transformer cores
Ferrites Mn- Zn ferrite	10^{-5}	0.4	<0.01	2000	5000	<0.01	HF low loss applications. HF transformers, recording heads, inductors (E and U cores)

5.5 Fe-Ni Alloys

Pure iron despite being soft magnetic material is not normally used as magnetic material because its good conductivity allows large eddy currents to be induced under varying fields. So, other materials are added to obtain required properties of the magnetic material. By adding (2-4)% of silicon to iron, silicon steel is obtained. Such material obtained is still soft magnetic material with increased resistivity but reduced eddy current loss in the machinery using this material. Silicon steel is most widely used in manufacture of power transformers and other electrical machinery for alternating current applications.

Nickel-iron alloys with composition around 77% Ni and 23% Fe constitute a soft magnetic material with low coercivity, low hysteresis loss and high initial and maximum relative permeabilities. High initial relative permeability makes the material useful in low magnetic field applications that are typically found in high frequency works in electronics e.g. audio and wide band transformers. They have found many engineering applications in the form of sensitive relays, pulse and wide band transformers, current transformers, magnetic recording heads, magnetic shielding, etc. Alloying iron with nickel increases resistivity and hence reduces the eddy current losses in the machinery. The domain wall motions are easy in nickel composition resulting smaller hysteresis losses in the material. There are number of commercial Ni-Fe alloys whose application depends on exact composition (which may also have a few percentages of Mo, Cr, Cu, etc) and method of preparation (e.g. mechanical rolling).

Amorphous magnetic materials have no crystalline structure they only have short-range orders. So they do not possess crystal imperfections such as grain boundaries and dislocations. They are prepared by rapid solidification of melt by using special techniques. Typically they are thin ribbons by virtue of their preparation method. Since there is no crystalline structure, all the directions are easy. The absence of crystalline defects, which normally impedes domain wall motions, leads to low coercivity and hence soft magnetic properties. The coercivity is not zero due to the directional nature of strains frozen in metal during rapid solidification. By virtue of their disordered structure, amorphous materials have high resistivity, and hence smaller eddy current losses. Although they are ideally suited for various transformers and electric machinery, their limited sizes and shapes prevent their use in power applications.

5.6 Ferrites for High Frequency Transformers

Ferrites are ferromagnetic materials, which are typically oxides of mixed transition metals, one of which is iron. Manganese ferrite is $MnFe_2O_4$, Manganese zinc ferrite is $Mn_{1-x}Zn_xFe_2O_4$ where x is less than unity. Ferrites are normally insulators, so they do not suffer from eddy current losses (very high resistance of the material allows only very small current and consequently very small or no losses). This is the reason for these being suitable for high frequency applications where eddy current losses given by $K_e f^2 B_m{}^2$ would prevent the use of material with reasonable conductivity.

Although ferrites can have high initial relative permeabilities and low losses, they do not possess as large saturation magnetization as ferromagnets. Their useful temperature range determined by Curie temperature is also lower. There are many types of ferrites available depending on the application, tolerable losses and required upper frequency of operation. Manganese zinc ferrites have high initial relative permeability of 2000 but are only useful up to a frequency of 1MHz, whereas Nickel zinc ferrites have lower initial relative permeability of 200 but they can be used up to a frequency of 200 MHz. Generally, the initial relative permeability in the high frequency region decreases with increase in frequency.

Garnets are ferrimagnetic materials that are used at highest frequencies covering the microwave range (1-300GHz). The yttrium iron garnet YIG ($Y_3Fe_5O_{12}$) is one of the simplest garnets with very low hysteresis loss at microwave frequencies. They have excellent dielectric properties with high resistivities and consequently low losses. The main disadvantage of garnets is the low saturation magnetization (0.18T for YIG) and low Curie temperature (280°C for YIG). The compositions of garnets depend upon the properties required for the particular microwave application. $Y_{2.1}Gd_{0.98}Fe_5O_{12}$ is a garnet that is used in 8-12GHz three-port circulators handling high microwave powers.

5.7 Hard Magnetic Materials

As we already mentioned they are hard to magnetize and demagnetize, and hence require large magnetizing field for both magnetization and demagnetization. The *B-H* curves of hard magnetic materials are broad and almost rectangular in shape. These materials possess relatively larger corcivities, which means they require large field to demagnetize them. The coercive field for hard magnetic materials can be millions of times greater than those for soft magnetic materials. Their characteristics make the hard magnetic materials suitable for making permanent magnets. It is clear that magnetization can be switched from one very persistent direction to other very persistent direction by a suitably large magnetizing field intensity. As the coercivity is strong, both the states $+B_r$ and $-B_r$ persist until a suitable large magnetic field intensity switches the field from one direction to the other. Hard magnetic materials can be used in magnetic storage of digital data where the states $+B_r$ and $-B_r$ can be made to represent 1 or 0 and vice versa.

An ideal hard magnetic material has a very large coercivity and remanent magnetic field. Since hard magnetic materials are used as permanent magnets, the energy stored per unit volume in external magnetic field should be as large as possible since this is the energy available to do work. This energy density in Joules per metre cube (Jm^{-3}) in external field depends upon the maximum value of the product of Band *H* in the second quadrant of *B-H* characteristics and is denoted by $(BH)_{max}$. This corresponds to the largest rectangular area that fits the *B-H* curve in the second quadrant.

When the size of a ferromagnetic sample falls below a certain critical dimension (0.1 micron for *Co*), the whole sample becomes a single domain because the energy required for generating a domain wall is too high compared to the reduction in external magnetostatic energy. These small particle-like pieces of magnetic materials are called single domain fine particles. Their magnetic properties depend both on the crystal structure and shape of the particle. For a spherical iron particle, the magnetization *M* will be in an easy direction e.g. along [001] in $+z$ direction. To reverse the magnetization from $+z$ to $-z$ by an external field, the magnetic spins have to be rotated around past the hard direction, since we can generate reverse domains. The rotation of magnetization involves substantial work due to magnetocrystalline anisotropy energy. So the sample has high coercivity. The higher the magnetocrystalline anisotropy energy, the greater is the coercivity. The energy involved in creating a domain wall increases with the magnetocrystalline anisotropy energy. The critical size below, which a particle becomes a single domain increases with magnetocrystalline anisotropy. Barium ferrite crystals have high degree of magnetocrystalline anisotropy due to their hexagonal shape. Critical size for single domain barium ferrite particles is about 1-1.5micron, and the coercivity about 0.3T compared with the coercivity of 0.02-0.1T for multi-domain barium ferrites.

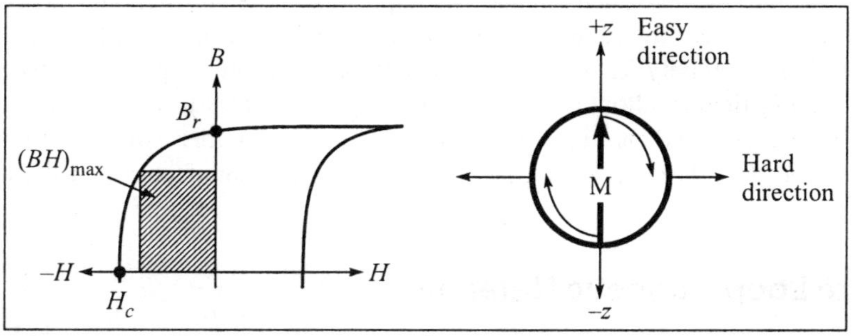

Fig. 5.16 $(BH)_{max}$ in hard magnetic material (a). A single domain fine particle (b)

Particles that are not spherical may even have higher coercivity as a result of shape anisotropy. For the single domain elongated particle, the potential energy in the external field is less for magnetization along long axis than for magnetization along short axis. So to rotate the magnetization from long axis to short, work has to be done. The magnetization of single domain fine particles along its elongated axis is termed as shape anisotropy. The elongation of the particle increases the coercivity and vice versa. Small spherical Fe-Cr-Co particles have a coercivity of 0.02T normally, but the elongated particles can have a coercivity as high as 0.075T due to shape anisotropy.

High coercivity magnets are fabricated by having elongated fine particles dispersed by precipitation in a structure. Alnico is one of the widely used permanent magnet material which is the alloy of aluminum, nickel, cobalt and iron. Fe-Co rich particles are dispersed in a matrix that is Al-Ni rich. The structure is obtained by appropriate heat treatment to allow Fe-Co particles to precipitate out from a solid solution of the alloy. Fe-Co particles possess strong magnetic field, whereas Al-Ni particles possess weak magnetic field. When heat treatment is carried out in the presence of strong magnetic field, the Fe-Co rich particles have their magnetization along elongated side. To demagnetize the elongated particles, high coercive field is required due to shape anisotropy. For manufacture of permanent magnets in permanent magnet machines, Alnico is widely used. The main disadvantage of this alloy is that it is mechanically hard and brittle. So, it cannot be shaped except by casting or sintering before heat treatment.

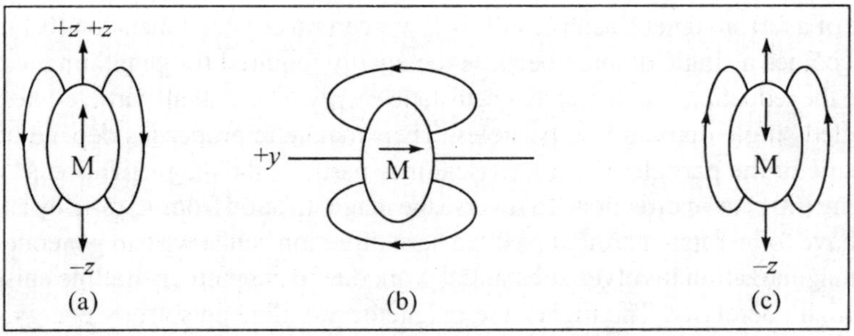

Fig. 5.17 Single domain elongated particle. Magnetization being changed from *a* to *b* to *c* by external field.

A variety of permanent magnets are made by compacting high coercivity particles by using powder metallurgy. The particles are magnetically hard because they are sufficiently small for each particle to be a single domain. Ceramic magnet, which is made by compacting barium ferrite, $BaFe_{12}O_9$ is very popular as permanent magnet. The barium ferrite has hexagonal crystal structure with large magnetocrystalline anisotropy giving high coercivity. These magnets are formed by wet pressing ferrite powder in the presence of strong magnetizing field to allow the single domain fine particles to be aligned in one of the easy directions, and then drying and carefully sintering the ceramic. Another example of ceramic magnet is strontium ferrite $SrFe_{12}O_9$. The ceramic magnets are mainly used to manufacture permanent magnets in low cost applications.

5.8 Square Loop Magnetic Materials

Rare earth cobalt permanent magnets based on samarium-cobalt (Sm-Co) alloys, which may possess other transition metals as well, have very high $(BH)_{max}$ values and are used in many applications such as stepper motors, servo motors, traveling wave tubes, gyroscopes, etc. $SmCo_5$ has a hexagonal crystal structure with high large magnetocrystalline anisotropy and hence high coercivity. $SmCo_5$ powder is pressed in the presence of strong magnetic field and then by carefully sintering solid powder magnet is formed. The $SmCo_{15}$ magnets are more recent and have $(BH)_{max}$ values of around 240kJm^{-3}.

The more recent neodymium-iron-boron, NdFeB, powdered solid magnets can have very large $(BH)_{max}$ values up to 275kJm^{-3}. The tetragonal crystal structure has the easy direction along long axis and possesses high magnetocrystalline anisotropy. Substantial amount of work has to be done to rotate magnetization around through the hard direction. These materials with very high coercivities are also termed as square loop magnetic materials because of the *B-H* loop imitating the square shape. To store any information in the form of digitalized data, these materials are most suited. The only drawback of these materials are lower Curie temperature compared to other magnetic materials such as Alnico with Curie temperature of 700°C compare to 300°C for NdFeB.

Table 5.2 Occurrence of permanent atomic magnetic dipoles and interaction between neighboring dipoles

Type of material	*Permanent dipoles*	*Interaction between neighboring dipoles*
Paramagnetic	Yes	Negligible
Ferromagnetic	Yes	Parallel orientation
Antiferrimagnetic	Yes	Antiparallel orientation of equal moments
Ferrimagnetic	Yes	Antiparallel orientation of unequal moments

Table 5.3 Susceptibilities of some paramagnetic materials at 300K

Substance	$\chi = \mu_r - 1, x10^{-3}$	*Substance*	$\chi = \mu_r - 1, x10^{-3}$
$CrCl_2$	1.5	Fe_2O_3	1.4
Cr_2O_3	1.7	$Fe_2(SO_4)_3$	2.2
CoO	5.8	$FeCl_2$	3.7
$CoSO_4$	2.0	$FeSO_4$	2.8
$MnSO_4$	3.6	$NiSO_4$	1.2

Table 5.4 Nèel temperature (T_N) and paramagnetic Curie temperature (θ_c) for some antiferrimagnetic materials.

Material	T_N, *K*	θ_c, *K*
MnF_2	72	113
MnO_2	84	316
MnO	122	610
MnS	165	528
FeO	198	570
NiF_2	73	116
CoO	292	280

Table 5.5 Susceptibility of some diamagnetic materials

Material	$\chi = \mu_r - 1, x10^{-5}$	*Material*	$\chi = \mu_r - 1, x10^{-5}$
Al_2O_3	-0.5	Cu	-0.9
$BaCl_2$	-2.0	Au	-3.6
NaCl	-1.2	Ge	-0.8
Diamond	-2.1	Si	-0.3
Graphite	-12	Se	-1.7

5.9 Solved Numerical Examples

1. Calculate the permeability and susceptibility of an iron bar of cross-sectional area 0.2cm² when a magnetizing field of 1200A. m⁻¹ produces magnetic field of 24 micro Weber.

Solution:

We have,

$$\phi = 2.4\times 10^{-5}\text{ Wb}$$
$$A = 0.2\text{cm}^2 = 0.2\times 10^{-4}\text{m}^2$$
$$H = 1200\text{A.m}^{-1}$$

Using the relation,

$$B = \frac{\phi}{A} = \frac{2.4\times 10^{-5}}{0.2\times 10^{-4}} = 1.2\text{T}$$

Now, the permeability is given by

$$\mu = \frac{B}{H} = \frac{1.2}{1200} = 10^{-3}\text{ H/m}$$

The susceptibility is then given by

$$\chi_m = \mu_r - 1 = \frac{\mu}{\mu_0} - 1 = \frac{10^{-3}}{4\pi\times 10^{-7}} - 1 = 737$$

2. Calculate the magnetization and flux density in a diamagnetic sample having susceptibility –0.3×10⁻⁵ and magnetic field strength 1000A/m.

Solution:

Given,

$$\chi = -0.3\times 10-5$$
$$H = 1000\text{ A/m}$$

We know that

$$M = \chi H = -0.3\times 10^{-5}\times 1000 = -310^{-3}\text{ A/m}$$

Now, the magnetic flux density is given by

$$B = \mu_0 H = 4\pi\times 10^{-7}\times 1000 = 1.257\times 10^{-3}\text{ T}$$

3. Calculate the magnetic moment of a bar of iron having length 0.5m and cross-sectional area 2mm² when placed in a long solenoid of 25 turns/cm carrying a current of 2A. Take the relative permeability of the material 400.

Solution:

Given, Length of the iron bar, l = 0.5m

Cross-sectional area, $A = 2\text{mm}^2 = 2\times 10^{-6}\text{m}^2$

No. of turns/length of the solenoid, $n = 25/\text{cm} = 25/10^{-2}\text{m} = 2500/\text{m}$

Current through the solenoid, I = 2A

We know that the magnetic field on the iron bar due to a solenoid is

$$B = \mu_0 \mu_r nI = 4\pi \times 10^{-7} \times 400 \times 2500 \times 2 = 2.51 \text{ T}$$

Now,

$$M = \frac{B}{\mu_0} = \frac{2.51}{4\pi \times 10^{-7}} = 1.998 \times 10^6 \text{ A.m}^{-1}$$

Hence the magnetic moment is equal to

$$= MV = MAI$$

$$= 1.998 \times 10^6 \times 2 \times 10^6 \times 0.5$$

$$= 1.998 \text{ Am}^2$$

4. The saturation induction of Nickel is 0.65Wb.m^{-2}. If the density is 8906Kg.m^{-3} and its atomic weight is 58.7; calculate the magnetic moment of Nickel atom in Bohr mangeton.

Solution:

We have, Saturation magnetic field, $B_s = 0.65 \text{Wb.m}^{-2}$

Density of the Nickel, $\rho = 8906 \text{Kg.m}^{-3}$

Atomic weight of Nickel, $A = 58.7\text{g}$

Magnetic moment for ferromagnetic materials is given by

$$\mu = \frac{\text{Magnetization}}{\text{No. of atoms/ volume}} = \frac{M}{n}$$

where

$$n = \frac{\rho N_A}{M} = \frac{8906 \times 6.02 \times 10^{23}}{58.7 \times 10^{-3}} = 9.13 \times 10^{28} \text{ m}^{-3}$$

$$\mu = \frac{B_S}{\mu_0 n} \text{ where } M = \frac{B_S}{\mu_0}$$

$$= \frac{0.65}{4\pi \times 10^{-7} \times 9.13 \times 10^{28}}$$

$$= 5.66 \times 10^{-24} \text{ A.m}^2$$

$$= 0.61 \mu_B$$

where $\mu_B = 9.273 \times 10^{-24}$ A.m^2 is Bohr magneton.

Exercise

1. Determine the permeability and relative permeability of a diamagnetic substance having susceptibility -9.5×10^{-9} [1.25×10^{-6}H/m; 0.99]
2. Calculate the relative permeability of a paramagnetic material at –73°C and 227°C if the susceptibility of the paramagnetic material at 27°C is 3.7×10^{-3}. [5.56×10^{-3} and 2.22×10^{-3}]

3. If a piece of ferric oxide has susceptibility 1.5×10^{-3} and a magnetic field intensity 10^6 A/m, then what are the magnetic flux density and magnetization of the material? [1.2T and 1500A.m^{-1}]
4. If an atom of a magnetic material consists of 10 electrons distributed uniformly and revolving round a circular path of radius 10^{-11}m, then calculate orbital dipole moment of the atom in Bohr magneton. [1.74 μ_B]
5. The magnetic field at northern magnetic pole of earth is vertical and is 6.2×10^{-5}T. Assuming the earth's magnetic field to be a dipole field and mean radius of the earth 6.4×10^6m, calculate (i) the earth's magnetic moment and (ii) the amount of current flowing round the single turn of wire round the equator to produce the dipole moment. [(i). 8.1×10^{22}A.m^2 (ii) 6.3×10^8A]
6. Calculate the field intensity of a magnetic field and the intensity of magnetization if 0.2A current is passed through a winding of 20 turns/cm over an iron anchor ring having magnetic field 1.26T. [3.98×10^2A.m^{-1} and 10^6A.m^{-1}]
7. The dipole moment associated with an atom of iron in an iron bar is 2.1×10^{-23}J/T. Assume that all the atoms in the bar, which is 5cm long and has a cross-sectional area of 10^{-4}m^2, have their dipole moments aligned. Calculate the dipole moment of the bar. [8.9Am^2]
8. A copper wire has 10^{29} electrons/m^3, cross-sectional area of 2mm^2 and carries current of 5A. Calculate the force acting on each electron if the wire is placed in a magnetic flux density 0.15T which is perpendicular to the wire. [3.75×10^{-24}N]
9. Calculate the current necessary to produce magnetic flux of 5×10^{-4}Wb in a soft iron ring of cross-sectional diameter 8cm, mean circumference 200cm with 400 turns of wire and relative permeability 1800. [0.22A]

Semiconductor Materials

Chapter 6

6.1 Band Structure of Group IV Materials, Energy Gap

Semiconductors in the present day world are one of the most important materials owing to their widespread use. Almost all the modern sophisticated engineering creations employ semiconductor devices for their effective functioning. XX century has witnessed phenomenal growth and advancement in semiconductor technology.

The group IV materials in Mendleev's Periodic Table such as Si and Ge have four electrons in the outermost orbit. These electrons are called valence electrons. To fulfill the outermost orbit each atom shares an electron with other atom and the result is the covalent bond. Not all the group IV elements are semiconductor in nature. Si and Ge in particular are semiconductors by their own in-built characteristics. They are called intrinsic semiconductors.

6.1.1 Energy Band in Silicon Crystal

The electronic configuration of Si is [Ne] $3s^2 3p^2$. However in the vicinity of other atoms, the 3s and 3p energy levels are so close that their interactions result in four new hybrid orbitals ψ_{hyb} that are symmetrically directed away from each other. Each hybrid orbital of one Si atom has one electron and half empty. So, hybrid orbitals of neighbouring atoms overlap and covalent bond with spin-paired electron is formed. A simplified two-dimensional illustration of bonding in Si is shown in figure 6.1.

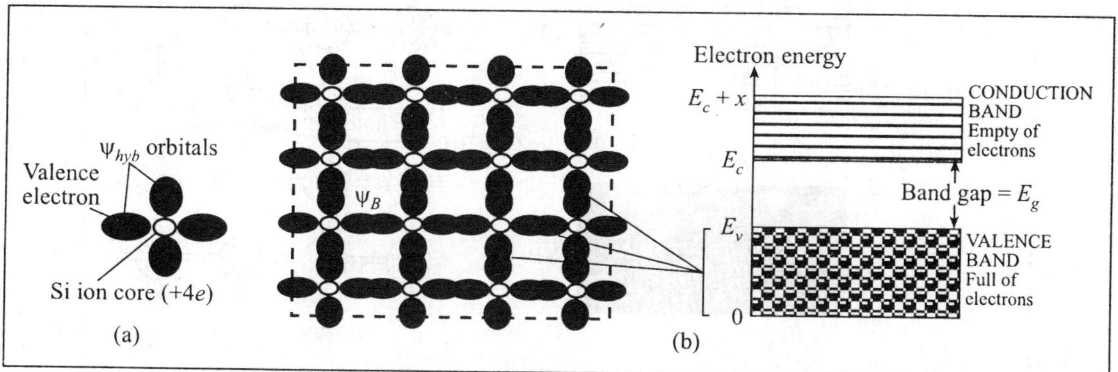

Fig. 6.1 Two dimensional sketch of bonding in Silicon crystal (a); Energy band diagram at 0K (b).

The conduction band contains electronic states that are at higher energies whereas the valence band contains electronic states that correspond to lower energies. The conduction band is separated by valence band by an energy gap E_g, called band gap. The width of the conduction band is called electron affinity and denoted by χ. The energy level E_c represents the bottom of the conduction band and E_v represents top of the valence band. An electron in the valence band like one in conduction band is not localized to an atomic site but extends throughout the whole solid. In the crystals, electrons can always interchange places by tunneling from one bond to the other. The wavefunction of a valence electron extends throughout the solid. So, the electrons in covalent bonds are indistinguishable. The silicon crystal can be crudely represented in as shown in figure 6.2.

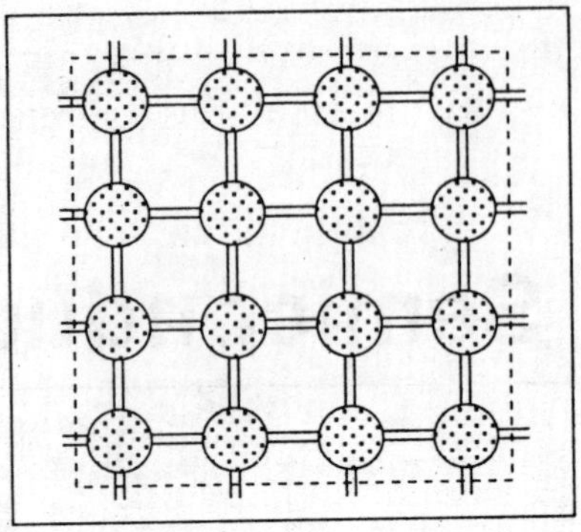

Fig. 6.2 Two-dimensional representation of Silicon crystal showing covalent bonds where each line represents a valence electron.

6.1.2 Electron and Hole

The covalent bond is saturated and there are no empty states available at valence band at absolute zero temperature. The conduction band is completely empty. An electron excited to conduction band is free to move around the crystal and can also respond to an applied filed because of the vacant energy levels around. At absolute zero, the valence electron cannot be excited to the conduction band. In order to overcome the energy gap between the two energy bands, electron at top of the valence band must gain an energy at least equal to the band gap, E_g. When a photon with energy greater than or equal to E_g is incident on the electron, it will be able to surmount the energy gap and find itself in the conduction band. The electron in the conduction band with its ability to move freely throughout the solid is termed as ***free electron***. The space left behind in the valence band by electron excited to the conduction band is called ***hole***. So, in intrinsic semiconductor (the materials are semiconductor on their own), only a free electron or

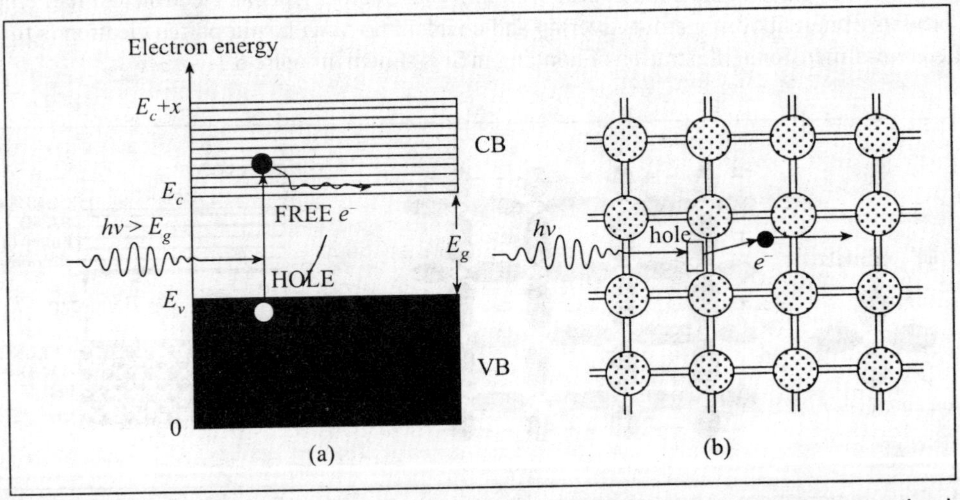

Fig. 6.3 Photon excites electron to conduction band (a). Electron-hole pair generation by incident photon (b).

conduction electron can not be generated but the generation of a conduction electron is inevitably accompanied by the generation of a valence hole. Such generated conduction electron and valence hole are together called ***electron-hole pair***.

Electron-hole pair can be generated not only by incident photons, but also by thermal means. Thermal energy causes lattice vibration in crystal, which will periodically deform the Si-Si bond, and in certain region, the atoms may be moving in such a way that the bond is overstretched and rupturing of the overstretched bond takes place. As the bond ruptures, the electron corresponding to that bond becomes free and is excited to conduction band. The empty electronic state with missing electron becomes positively charged, as there is shortage of one electron. This empty space with positive charge is denoted by h^+ and can also move throughout the whole valence band in the solid in a manner similar to the conduction electron. As we see, in figure 6.4, the neighbouring electron can tunnel through to the vacant place and the hole or vacant place is moved in the direction opposite to the direction of movement of electron in valence band. So, the hole moves in the crystal in the manner if it were a free positively charged particle. When an electric field is applied, the holes will drift in the direction of applied field and contribute to the electric current in the manner similar to the current due to conduction electrons. The current due to both the conduction electron and valence hole is in the same direction. Hole is an empty electronic state in valence band, which behaves as if it were a positively charged particle that is free to respond to an applied electric field.

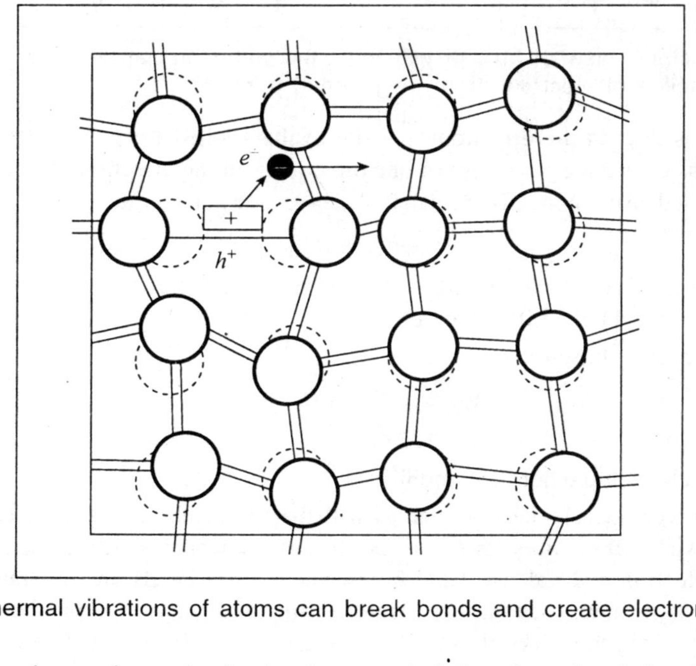

Fig. 6.4 Thermal vibrations of atoms can break bonds and create electron-hole pair.

When a wandering electron in conduction band meets a hole in valence band, the electron will occupy the empty state, which is at lower energy than the present energy state of electron. The electron falls from conduction band to valence band and fills the hole. This process is called recombination (figure 6.5e).

6.1.3 Electrical Conduction in Semiconductor

When an electric field is applied to the semiconductor, both the electrons in the conduction band and holes in the valence band start drifting. Valence holes drift in the direction of applied field whereas

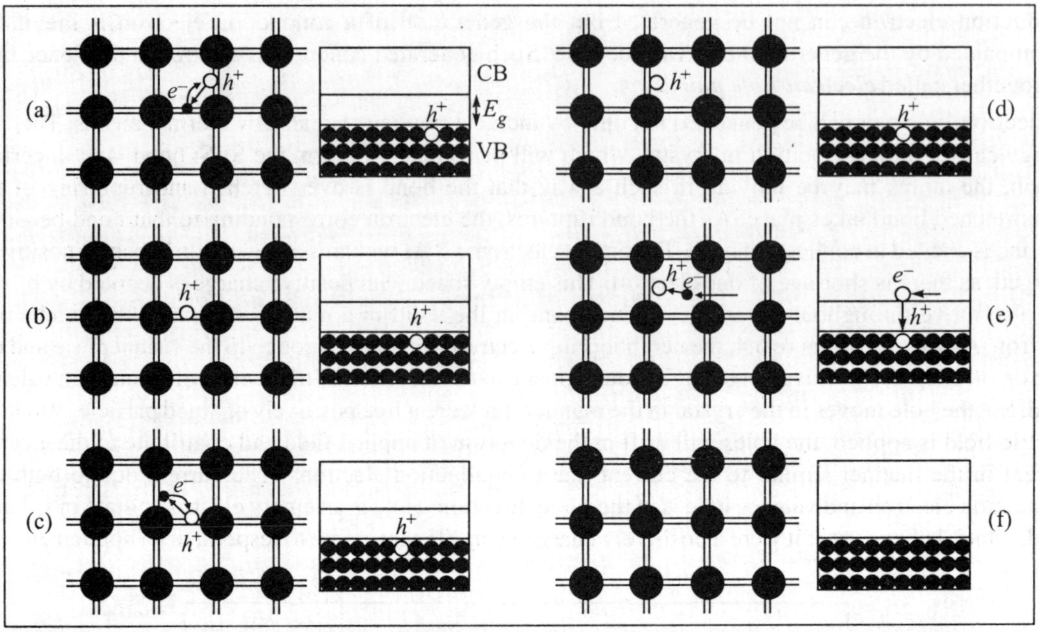

Fig. 6.5 A pictorial view of hole in valence band wandering around the crystal due to tunneling of electrons from neighboring bonds.

conduction electrons drift in the direction opposite to the applied field. Both the valence holes and conduction electrons contribute to current in one direction – in the direction of applied field. From the expression of current density in metals, we can make an analogy and write

$$J = env_{de} + epv_{dh} \tag{6.1}$$

where n and p are respective concentrations of electrons and holes in conduction band and valence band; v_{de} and v_{dh} are electron and hole drift velocities.

The drift velocities are (in x-direction)

$$v_{de} = \mu_e E_x$$
$$v_{dh} = \mu_h E_x \tag{6.2}$$

where μ_e and μ_h are electron and hole drift mobilities.

The drift mobilities can be calculated in analogy with that for metals in chapter II. The only difference is in that case we derived drift mobility for free electron in metal, whereas, here these are the mobilities for conduction band electron and valence band holes where these bands are separated by considerable energy gap. In case of metal we used mass of a free electron to calculate mobility, but here we have to use the effective mass of electron in crystal to account for the internal forces that hinders its movement. For hole also we will use its effective mass, which is different from effective mass of electron in co induction band.

$$\mu_e = \frac{e\tau_e}{m_e^*} \tag{6.3}$$

And,
$$\mu_h = \frac{e\tau_h}{m_h^*} \tag{6.4}$$

The conductivity of the semiconductor is given by

$$\sigma = en\mu_e + en\mu_h \tag{6.5}$$

6.2 Electron and Hole Concentrations in Intrinsic Semiconductor

The conductivity of semiconductor is given by equation (6.5) and mainly depends up on the electron and hole concentrations in conduction and valence bands respectively. By using the Fermi-Dirac function, we will calculate the number of electrons and holes per unit volume in semiconductor.

If n_E is the number of electrons per unit energy per unit volume, $Z(E)$ be the density of states in conduction band, $F(E)$ is the probability of occupation, then the number of electrons in the energy interval E to $E + dE$ is given by

$$n_E dE = Z(E)\, F(E)\, dE \tag{6.6}$$

Integrating the above expression from bottom of the conduction band, E_c to top of conduction band, $E_c + \chi$, we will find the electron concentration or number of electrons per unit volume in conduction band.

$$n_i = \int_{E_c}^{E_c+\chi} n_E dE = \int_{E_c}^{E_c+\chi} Z(E)\, F(E)\, dE \tag{6.7}$$

Assuming that $(E_c - E_F) >> kT$, we can write

$$F(E) \approx \exp\left[-\frac{E - E_F}{kT}\right] \tag{6.8}$$

which is Boltzmann's statistics.

The upper limit in the integral will be $E = \infty$ rather than $E_c + \chi$ as $F(E)$ decays rapidly with energy so that $Z(E)\, F(E)$ tends to zero near the top of the band. $Z(E)$ is significant only close to the bottom of conduction band.

$$Z(E) = \frac{8\sqrt{2}\pi(m_e^*)^{3/2}}{h^3}(E - E_c)^{1/2} \tag{6.9}$$

Putting from equation (6.9) into equation (6.7), we get

$$n = \frac{8\sqrt{2}\pi\left(m_e^*\right)^{\frac{3}{2}}}{h^3}\int_{E_c}^{\infty}(E - E_c)^{\frac{1}{2}}\exp\left[-\frac{E - E_F}{kT}\right]$$

Let us make a substitution $\frac{E - E_C}{kT} = x$.

So that $dE = (kT)\, dx$

And the above expression becomes

$$n = 4\pi\left(\frac{2kTm_e^*}{h^2}\right)^{\frac{3}{2}}\exp\left(-\frac{E_c - E_F}{kT}\right)\int_{E_0}^{\infty}(x)^{\frac{1}{2}}\exp(-x)\, dx \tag{6.10}$$

But, we have from table of integrals

$$\int_0^\infty (x)^{1/2} \exp(-x)\, dx = \frac{\sqrt{\pi}}{2}$$

And the expression (6.10) becomes

$$n = 2\left(\frac{2\pi kT m_e^*}{h^2}\right)^{\frac{3}{2}} \exp\left[-\frac{E_c - E_F}{kT}\right]$$

or

$$n = N_c \exp\left[-\frac{E_c - E_F}{kT}\right] \tag{6.11}$$

where $N_c = 2\left[\frac{2\pi kT m_e^*}{h^2}\right]^{3/2}$ is called effective density of states at the conduction band edge.

So, by multiplying effective density of states at conduction band edge by Boltzmann's statistics, we can calculate the electron concentration at E_C.

In the similar manner, we can obtain the concentration of holes at the valence band. The probability of occupancy of hole can be found by subtracting the probability of occupation function from unity, indicating the absence of electron. Integrating over valence band, we obtain

$$p = \int_0^{E_v} p_E dE = \int_0^{E_v} Z_{VB}(E)\,[1 - F(E)]\, dE \tag{6.12}$$

As in the previous case, assuming that E_V is few kT above the top of valence band, and performing the integration, we will find

$$p \approx N_v \exp\left[-\frac{E_F - E_v}{kT}\right] \tag{6.13}$$

where N_V is called effective density of states at the valence band edge and equal to

$$N_v = 2\frac{(2\pi m_h^* kT)^{3/2}}{h^3} \tag{6.14}$$

Equations (6.10) and (6.13) give the electron concentration in conduction band and hole concentration in valence band.

Multiplying the hole concentration and electron concentration, we will have

$$np = N_c \exp\left[-\frac{E_c - E_F}{kT}\right] N_v \exp\left[-\frac{E_F - E_v}{kT}\right]$$

$$np = N_c N_v \exp\left[-\frac{E_c - E_v}{kT}\right] \tag{6.15}$$

Equation (6.15) is commonly termed as "Mass action law".

But $\quad E_c - E_v = E_g$

So
$$np = N_c N_v \exp\left[-\frac{E_g}{kT}\right] \tag{6.16}$$

For intrinsic semiconductors, the electron and hole are generated simultaneously. This means their concentration must be same. The product of ***np*** is constant for a given temperature.

$$n = p = n_i$$

where n_i represents intrinsic concentration of the semiconductor.

The intrinsic concentration can be found as

$$n_i = (N_c N_v)^{1/2} \exp\left[-\frac{E_g}{2kT}\right] \tag{6.17}$$

Since

$$n = p = n_i$$

or,
$$N_v \exp\left[-\frac{E_F - E_v}{kT}\right] = (N_c N_v)^{1/2} \exp\left[-\frac{E_g}{2kT}\right] \tag{6.18}$$

For intrinsic semiconductor, the Fermi energy level is denoted by E_{Fi}, which can be found from the above relationship.

$$\exp\left[-\frac{E_{Fi} - E_v}{kT}\right] = \frac{1}{N_v} (N_c N_v)^{1/2} \exp\left[-\frac{E_g}{2kT}\right]$$

Taking natural logarithm on both the sides, we will have

or,
$$-\frac{E_{Fi} - E_v}{kT} = \ln(N_c/N_v)^{1/2} + \left[-\frac{E_g}{2kT}\right]$$

or,
$$-E_{Fi} + E_V = kT\ln(N_c/N_v)^{1/2} + kT\left[-\frac{E_g}{2kT}\right]$$

or,
$$E_{Fi} = E_v - \frac{1}{2} kT\ln(N_C/N_V) + \frac{E_g}{2}$$

or,
$$E_{Fi} = E_v - \frac{1}{2} kT\ln\frac{(8\pi m_e^* kT/h^2)^{3/2}}{(8\pi m_h^* kT/h^2)^{3/2}} + \frac{E_g}{2}$$

or,
$$E_{Fi} = E_v + \frac{E_g}{2} - \frac{3}{4} kT\ln\frac{m_e^*}{m_h^*} \tag{6.19}$$

In the above equation (6.19), if effective masses of electron in conduction band and of hole in valence band are equal to each other, the intrinsic Fermi level in semiconductor lies in the middle of the energy gap.

$$E_{Fi} = E_v + \frac{E_g}{2}$$

$$E_{Fi} = E_v + \frac{E_c - E_v}{2} = E_c - \frac{E_c - E_v}{2} = E_c - \frac{E_g}{2}$$

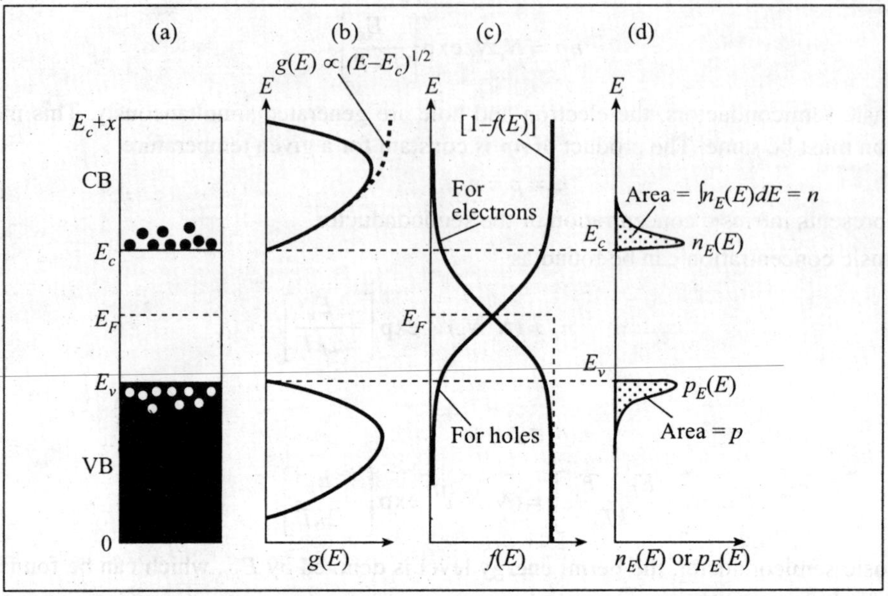

Fig. 6.6 Energy band (a), Density of states (b), Probability of occupation (c), electron or hole concentration (d).

or,
$$E_{Fi} = E_c - \frac{E_g}{2} \tag{6.20}$$

Normally, effective masses of conduction electron and valence hole differ slightly, which means the intrinsic Fermi level does not lay precisely at the center of the energy band gap, but is shifted a little bit. In Si, the hole effective mass is slightly greater than that of electron, so the intrinsic Fermi level is slightly above the mid-gap.

Some important properties of the important semiconductors are presented in table 6.1.

6.3 Extrinsic Semiconductors

Pure semiconductors do not always serve our purpose or give operating characteristics of semiconductor material according to our requirement. For this, pure semiconductors are added with materials belonging to either group III or group V in Mendeleev's periodic table. These group III or group V elements (materials)

Table 6.1 Different parameters of different semiconductors

Semi-conductor	E_g, *eV*	χ, *eV*	N_c, cm^{-3}	N_v, cm^{-3}	n_i, cm^{-3}	μ_e, cm^2 V^{-1} s^{-1}	μ_h cm^2 V^{-1} s^{-1}	m_e^*/m_e	m_h^*/m_e	ε_r
Ge	0.66	4.13	1.04 10^{19}	6 10^{18}	2.4 10^{13}	3900	1900	.12a 0.56b	0.23a 0.4b	16
Si	1.1	4.01	2.8 10^{19}	1.04 10^{19}	1.45 10^{10}	1350	450	0.26a 1.08b	0.38a 0.56b	11.9
GaAs	1.42	4.07	4.7 10^{17}	7 10^{18}	1.8 10^{6}	8500	400	0.067a,b	0.4a 0.5b	13.1

Note: Effective masses related to conductivity(a) and density of states (b).

added to the pure semiconductor (mostly the silicon) are called ***impurities***. And the resulting semiconductor is known as ***extrinsic semiconductor***. We can obtain different types of extrinsic semiconductors by adding different types of impurities to the pure semiconductor. There are four valence electrons in silicon. If pentavalent impurities like arsenic and phosphorous are added, four electrons will participate to form covalent bond with neighbouring silicon atoms, whereas the fifth electron cannot participate in bonding since all the bonds are saturated. This extra electron can be contributed to the conduction band. The impurity atom is donating an extra electron to the conduction band of the semiconductor. Such semiconductors are called n-type (remember *n* from word do***n***or). Similarly, we can get p-type (remember *p* from word acce***p***tor) semiconductor by adding trivalent impurities like boron, aluminum, gallium, etc.

6.3.1 n-type Semiconductor

By adding small amount of pentavalent impurities into the silicon, we can obtain wide range of characteristics of the resultant semiconductor. When As is added to Si crystal as shown in figure 6.7, it is forced to bond with neighbouring Si atoms in the same diamond crystal structure as Si as it is surrounded by millions of these silicon atoms. Only four of the five valence electrons in the arsenic will participate in bonding with neighbouring silicon atoms. The fifth has nothing to do in bond formation. This extra electron is continuously orbiting the As^+ ion core as shown in figure 6.7 similar to the hydrogen atom in silicon environment. The energy required to free the electron from As site and consequently ionizing As atom can be calculated in analogy with ionization energy of hydrogen, which means to free its electron from the ground state corresponding to $n = 1$. The binding energy of the electron in hydrogen atom is

$$E_b = -E_1 = \frac{e^4 m_e}{8\varepsilon_0^2 h^2} = 13.6 \text{ eV}$$

For electron around As^+ ion core, we have to use instead of absolute permittivity of air, permittivity of silicon and effective mass of electron in Si.

$$E_b^{Si} = \frac{e^4 m_e^*}{8\varepsilon_0^2 \varepsilon_r^2 h^2} = 0.032 \text{ eV}$$

where $m_e^* = m_e/3$ and $\varepsilon_r = 11.9$.

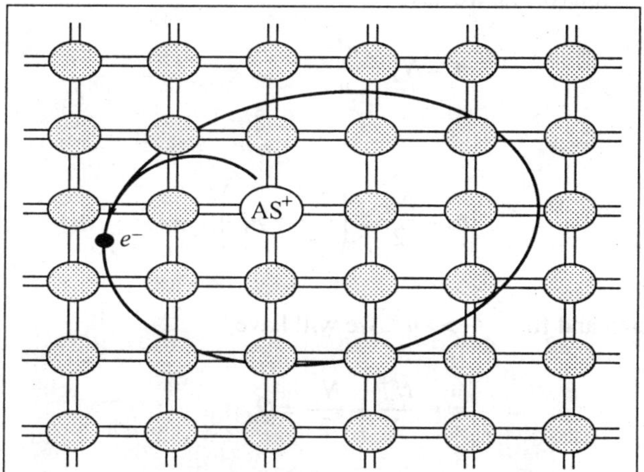

Fig. 6.7 Simplified two-dimensional pictorial view of As-doped Si crystal.

Average thermal energy of atomic vibration is ~3kT(~0.07eV). Thus the free fifth valence electron can readily be freed by thermal vibrations in the silicon lattice and excited to the conduction band. The introduction of pentavalent impurity like arsenic creates localized electronic state with localized hydrogenic wavefunction around As^+. In the energy band diagram, these donor sites are represented by E_d as shown above in figure 6.8.

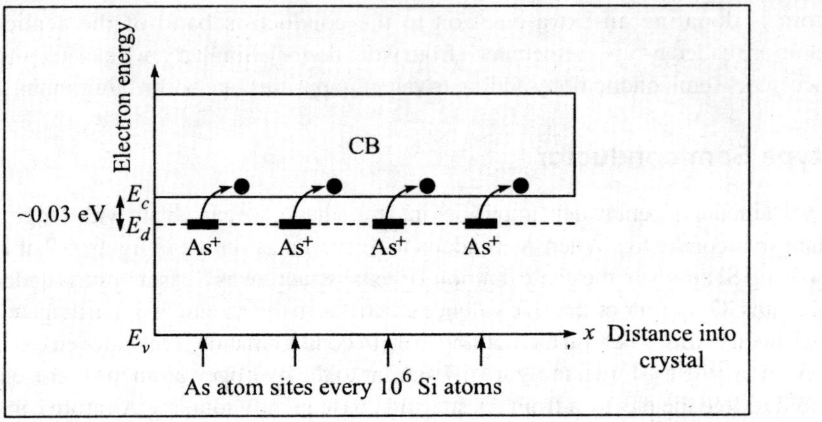

Fig. 6.8 Energy band diagram for an n-type semiconductor doped with arsenic.

In n-type semiconductor, the conduction electrons, which outnumber valence holes, are called ***majority charge carrier*** and the holes called ***minority charge carrier***. In equilibrium condition, a semiconductor is charge neutral, i.e, it possesses equal negative and positive charges. All the donor sites are easily ionized. If n and p represent the extrinsic electron and hole concentration of semiconductor, N_d the donor concentration, then supposing all donor sites are ionized (all donor atoms contribute one electron to CB), electron concentration is given by

$$n = N_d + p \tag{6.21}$$

But

$$np = n_i^2$$

$$p = n_i^2/n$$

Substituting for p in equation (6.21), we have

$$n = N_d + \frac{n_i^2}{n}$$

which gives

$$n = \frac{N_d}{2} \pm \left(\frac{N_d^2}{4} + n_i^2\right)^{\frac{1}{2}} \tag{6.22}$$

Ignoring the negative sign and for $\frac{N_d^2}{4} >> n_i^2$, we will have

$$n \approx \frac{N_d}{2} + \frac{N_d}{2} = N_d \tag{6.23}$$

Similarly, we can find the minority carrier concentration as

$$p \approx \frac{n_i^2}{N_d} \tag{6.24}$$

The hole concentration in n-type semiconductor given by equation (6.24) is less than the intrinsic hole concentration. This is because some of the electrons in conduction band recombine with holes in valence band so as to maintain $np = n_i^2$. As we go on increasing concentration of donor atoms thereby increasing the majority carrier concentration, the minority carrier concentration will be decreased simultaneously. This is called ***minority carrier suppression***. It is similar to if the minority carriers were suppressed by the majority carriers.

The conductivity of semiconductor is then

$$\sigma = en\mu_e + ep\mu_h = eN_d\mu_e + e\mu_h\left(\frac{n_i^2}{N_d}\right)$$

or,

$$\sigma \approx eN_d\mu_e \tag{6.25}$$

So, the conductivity is mainly due to the mobility of electrons in the n-type semiconductors.

6.3.2 p-type Semiconductor

As we already know, by adding trivalent impurities to the silicon, we get extrinsic semiconductor of p-type. When very few trivalent impurities (1 impurity atom in 1 mln Si atoms) are added to silicon, they are forced to occupy the same diamond crystalline structure as that of silicon – there is no alternative for the impurity atoms. But when doing so, the bond formed between impurity atom and surrounding four silicon atoms is incomplete because there are only three valence electrons in impurity atom, which can participate in bonding. The unfilled electronic state is the hole. By doping silicon with trivalent impurity like boron, we can create hole in the valence band without creating a conduction electron at the same time. So, we can create only one of the free charge carriers in semiconductor by adding suitable impurities, which is not the case in intrinsic semiconductor (electron and hole are simultaneously created).

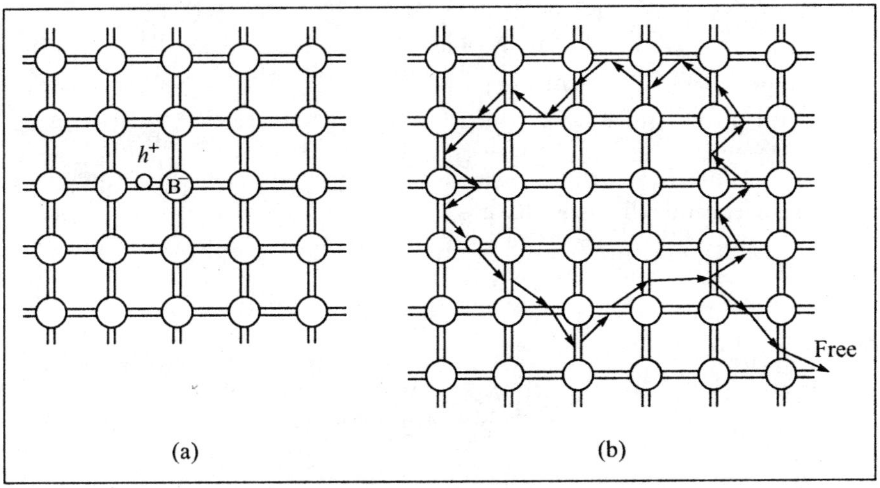

Fig. 6.9 Adding *B* to Si creates a hole (a).

The hole orbits around the B^- site by tunneling of electrons from neighboring Si atoms (b). Thermally vibrating Si atoms provide enough energy to free the hole from B^- site into the valence band.

As shown in the figure, as the hole moves away from negatively charged ion core, it is attracted by it. Consequently the hole orbits around the ion core. The binding energy is very small ~0.05 eV, so at room temperature all the acceptor sites are ionized by thermal vibrations in the lattice. When boron accepts electron from neighboring Si-Si bond, the hole is shifted away from the boron to Si-Si bond in valence band. The boron atom doped to silicon thus accepts an electron from Si-Si bond, and the resulting semiconductor is called p-type (remember p from acceptor).

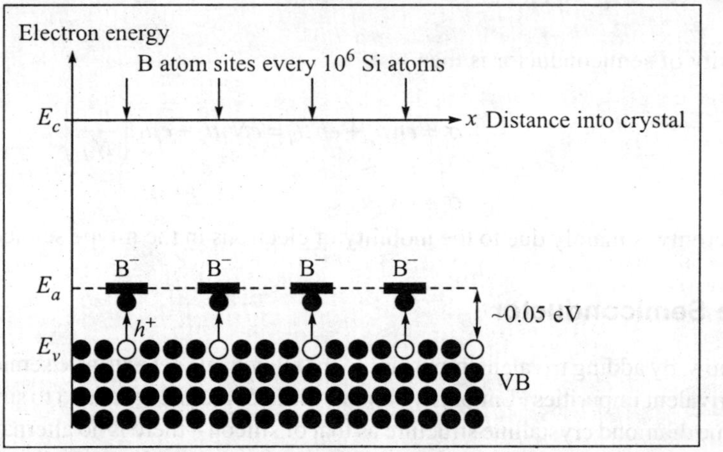

Fig. 6.10 Energy band diagram for p-type semiconductor doped with B showing the ionized acceptor sites and creation of hole in valence band.

Adding trivalent impurities will increase the hole concentration in valence band but at the same time does not increase the electron concentration in conduction band. If n and p be the electron and hole concentration in the semiconductor and N_a the acceptor concentration, then hole concentration is given by

$$p = N_a + n \tag{6.26}$$

But

$$np = n_i^2$$

$$n = \frac{n_i^2}{p}$$

Substituting for p in equation (6.26), we have

$$p = N_a + \frac{n_i^2}{p}$$

which gives

$$p = \frac{N_a}{2} \pm \left(\frac{N_a}{2} + n_i^2\right)^{1/2} \tag{6.27}$$

Ignoring the negative sign and for $\frac{N_a^2}{2} >> n_i^2$, we will have

$$p \approx \frac{N_a}{2} + \frac{N_a}{2} = N_a \tag{6.28}$$

The conductivity in p-type semiconductor is mainly due to the hole concentration which can be calculated as

$$\sigma = en\mu_e + ep\mu_n = e\mu_e\left(\frac{n_i^2}{N_a}\right) + eN_a\mu_h$$

or,

$$\sigma \approx eN_a\mu_n \tag{6.29}$$

Table 6.2 Examples of donor and acceptor ionization energies (eV) in silicon.

Donors			Acceptors		
Phosphorous	*Arsenic*	*Antimony*	*Boron*	*Aluminum*	*Gallium*
0.045	0.054	0.039	0.045	0.057	0.072

In trivalent doping of silicon, the electron concentration in hole is decreased as a result of some of the large number of holes created by doping recombine with electrons in conduction band thereby decreasing the electron concentration, which is in minority in p-type semiconductor. This means the minority charge carriers are suppressed by majority charge carriers. This phenomenon is called ***minority carrier suppression***.

6.3.3 Compensation Doping

As we have observed in previous discussions, the properties of semiconductors are changed by doping with either trivalent or pentavalent impurities. The level of doping can be different so as to obtain different properties and operating characteristics of the resulting semiconductor device. It is interesting to see what happens when both the trivalent and pentavalent impurities are added to silicon. Adding trivalent impurities will increase the hole concentration in the semiconductor whereas by adding pentavalent impurities to the same will compensate for the increase in hole concentration and decrease in electron concentration. Assuming all donor impurities and acceptor impurities ionized at room temperature, the effective hole concentration and electron concentration are calculated based on which of the impurity concentrations is greater. If N_a and N_d are donor concentration and acceptor concentration respectively, and $N_d > N_a$, the holes due to acceptor doping will recombine with the electrons due to donor doping. So, the electron concentration is

$$n = N_d - N_a$$

$$p = \frac{n_i^2}{n} = \frac{n_i^2}{N_d - N_a} \tag{6.30}$$

Similarly, when $N_a > N_d$, the electrons due to donor doping recombine with holes due to acceptor doping and the resultant concentrations are

$$p = N_a - N_d$$

$$n = \frac{n_i^2}{p} = \frac{n_i^2}{N_a - N_d} \tag{6.31}$$

where n_i is intrinsic concentration in the semiconductor.

6.4 Temperature Dependence of Carrier Concentration

Rewriting the expression (6.17) for intrinsic concentration of semiconductor

$$n_i = \sqrt{N_c N_v}\ \exp\left[-\frac{E_g}{2kT}\right]$$

where $n_i = n = p$ is valid for intrinsic semiconductors. Equation (6.17) implies that electron and hole concentrations in intrinsic semiconductors are the functions of energy band gap and temperature of the sample. Both the N_c and N_v have $T^{3/2}$ in their expressions, so they are also the functions of temperature. Since there is negative sign in exponential term, it signifies that the increase in band gap will reduce the intrinsic concentration and vice versa. Both from the exponential term and N_c and N_v it can be observed that the increase in temperature will cause increase in concentration as well. The energy band gap is the built-in characteristic of all materials, which cannot be altered, whereas the temperature of the sample can be controlled to suit our purpose. The intrinsic concentrations of some of the important semiconductors as function of temperature are presented in figure 6.11.

In extrinsic semiconductor, the carrier concentration depends on different parameters at different temperatures. For an n-type semiconductor, at very low temperature, the donor sites will not be ionized and carrier concentration will be very low and determined only by the intrinsic concentration at these temperatures. As temperature increases, thermal energy of lattice vibrations become more and more, and

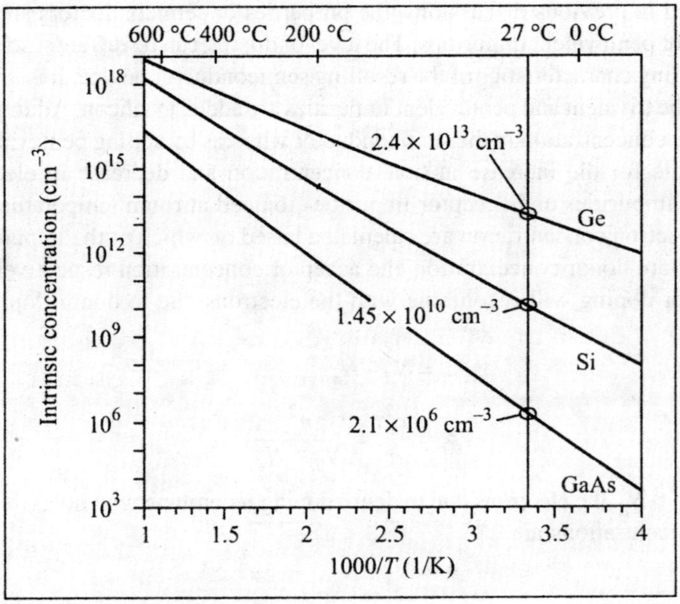

Fig. 6.11 Temperature dependence of carrier concentration of intrinsic semiconductors

most of the donor sites are ionized thereby defining the carrier concentration. In the high temperature range again, the generation of electron-hole pair in semiconductor is more than the impurity concentration, which means even though all impurity sites are ionized their contribution to carrier concentration is surpassed by the electron-hole concentration from within the semiconductor itself. So at high temperatures, doped semiconductor again behaves like intrinsic semiconductor.

In extrinsic semiconductor, the electron concentration at low temperature when thermal vibrations are unable to ionize donor sites and excite the electrons from donor site E_d to conduction band E_c can be expressed as

$$n = \sqrt{\frac{1}{2} N_c N_v} \exp\left[-\frac{E_c - E_d}{2kT}\right]$$

$$n = \sqrt{\frac{1}{2} N_c N_v} \exp\left[-\frac{\Delta E}{2kT}\right] \tag{6.32}$$

The ionization of donor atoms and electron concentration in conduction band at different temperatures in arsenic doped silicon semiconductor is shown in figure 6.12.

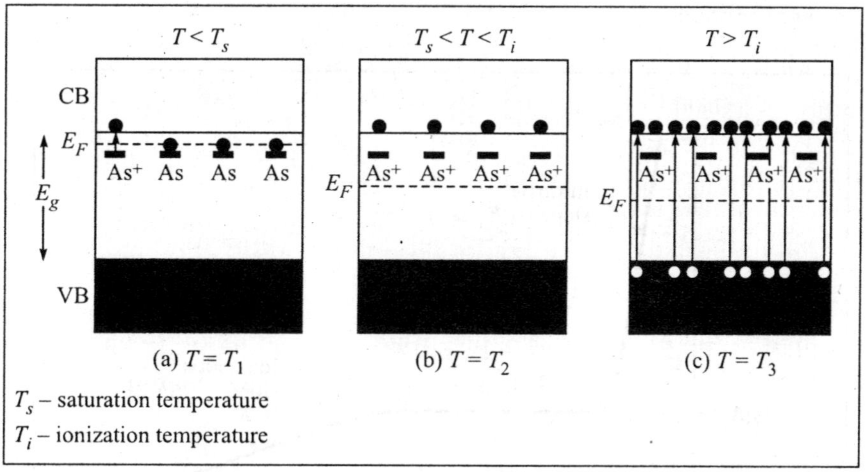

Fig. 6.12 Energy band diagram showing ionized donors at different temperatures

In the log scale, temperature dependence of electron concentration in n-type semiconductor will be as presented in figure 6.13 below.

6.5 Drift and Diffusion of Charge Carriers

6.5.1 Drift of Charge Carriers

The drifting of charge carriers under the influence of applied electric field is similar to that of electrons in metal. In semiconductor, in addition to drift of conduction electrons, there is drift of valence holes as well. The drift velocities of electron and hole are expressed in x-direction as (in analogy with drift velocity of electron in metal)

$$v_{de} = \mu_e E_x \tag{6.33}$$

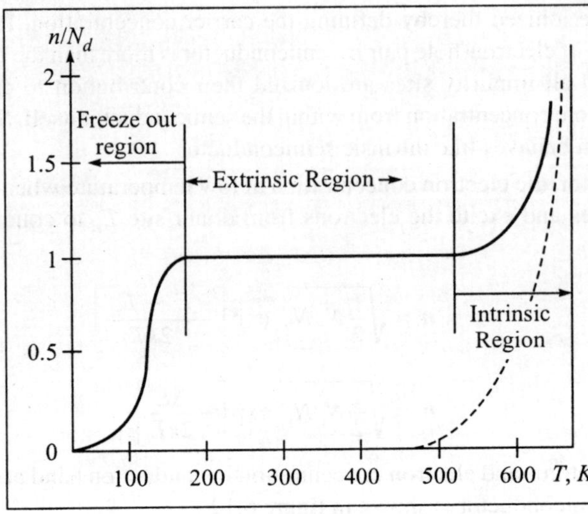

Fig. 6.13 Temperature dependence of extrinsic semiconductor in n-type silicon semiconductor

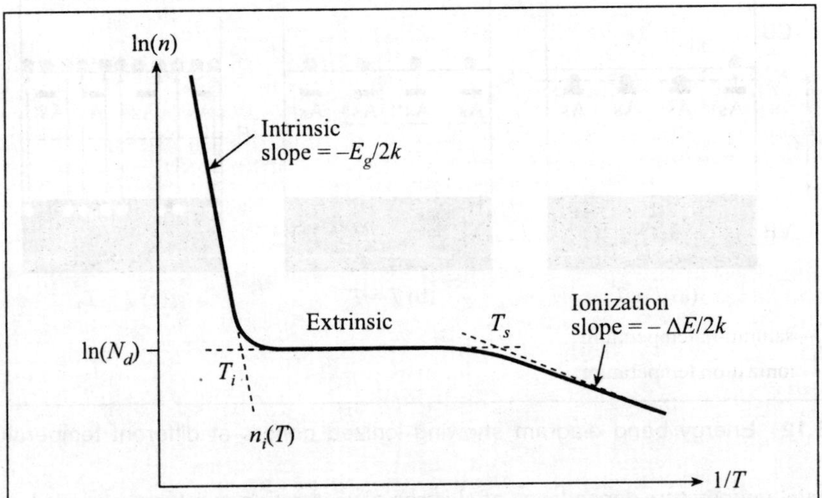

Fig. 6.14 Temperature dependence of electron concentration in n-type semiconductor

$$v_{dk} = \mu_h E_x \tag{6.34}$$

where E_x is the applied electric filed in x-direction; μ_e and μ_h are electron and hole mobilities respectively.

The current density due to the drift of conduction electrons and valence holes is given by

$$J_{drift} = \sigma E_x \tag{6.35}$$

where σ is the conductivity of semiconductor.

$$J_{drift} = (en\mu_e + ep\mu_h)\, E_x \tag{6.36}$$

Conductivity and hence drift current density both depend on the mobility of charge carrier. Mobility is the measure of ease with which the charge carrier can move within the semiconductor crystal. The charge carriers when drifting in the crystal constantly collide with lattice atoms and with each other, which reduces their mobility. This is the region why in highly doped semiconductors, the mobility is considerably reduced than in lightly doped one. Similarly, increase in temperature causes increase in lattice vibrations, which in turn will reduce the mobility of the carriers.

6.5.2 Temperature Dependence of Mobility

Recalling from the previous chapter, electron drift mobility in metal is given by

$$\mu = \frac{e\tau}{m_e^*}$$

where τ is the mean free time between scattering, which itself is given by

$$\tau = \frac{1}{Sv_{th}N_s} \tag{6.37}$$

where S – cross-sectional area of scatterer

v_{th} – mean speed of electrons, called thermal velocity;

N_s – number of scatterers per unit volume.

After series of mathematical manipulations, we will arrive at a relationship between mobility limited by lattice vibration scattering in the form

$$\mu_L \propto T^{-\frac{3}{2}} \tag{6.38}$$

And similarly, ionization scattering limited mobility will be

$$\mu_I \propto \frac{T^{\frac{3}{2}}}{N_I} \tag{6.39}$$

where N_I – is the concentration of ionized impurities. Expression (6.39) shows that with increase in temperature as all the impurities are ionized, the mobility limited by ionization scattering also decreases.

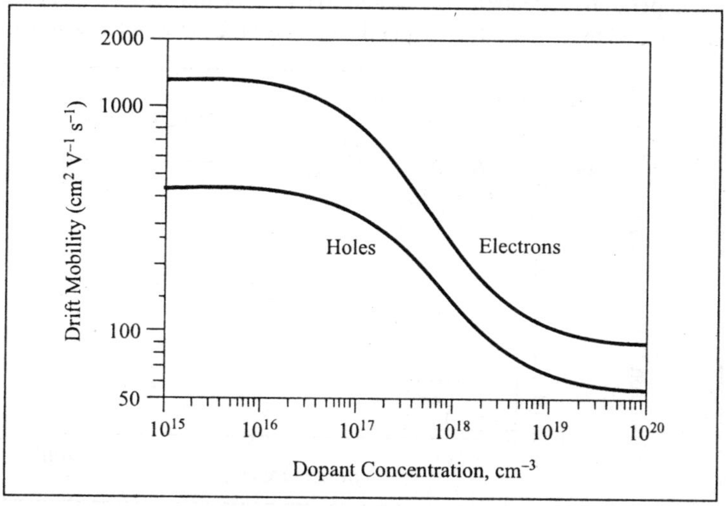

Fig. 6.15 Log-log plot of experimental values of drift mobility versus temperature for n-type semiconductors. The concentrations are in per cm^3.

Mobility of electrons and holes in n-type semiconductor at a doping concentration of ~ 10^{14} per cm^3 can be expressed as function of temperatures as $\mu_e \propto T^{-23}$ and $\mu_h \propto T^{-23}$

Mobility in extrinsic semiconductor heavily depends on the doping concentration. Assuming all impurities ionized at room temperature, the variation in drift mobility of charge carriers in extrinsic semiconductors will be as shown in figure 6.16.

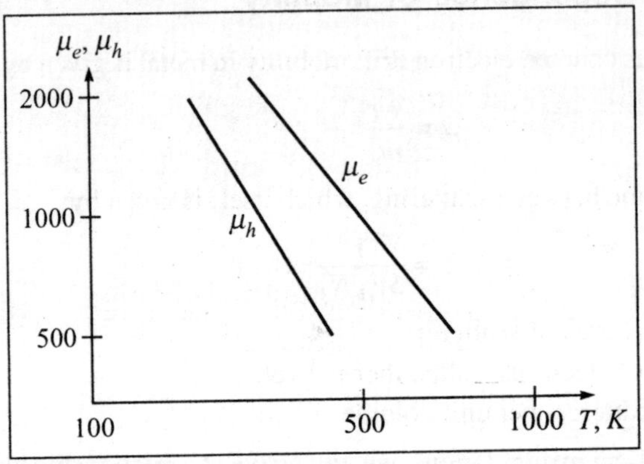

Fig. 6.16 Temperature dependence of mobility in extrinsic semiconductor

The conductivity of semiconductor as given by $\sigma = en\mu_e + ep\mu_h$ depends on carrier concentration and drift mobility. For n-type semiconductor (in extrinsic range), it will depend on donor concentration and electron drift mobility whereas in p-type – on acceptor concentration and hole drift mobility. In n-type semiconductor, at lowest temperatures in the ionization range, the electron concentration depends exponentially on the temperature.

$$n = \left(\frac{1}{2} N_c N_v\right)^{1/2} \exp\left[-\frac{E_c - E_d}{2kT}\right]$$

which dominates the temperature dependence of conductivity. In the intrinsic range at highest temperatures, the conductivity is dominated by temperature dependence of n_i since $\sigma = en_i(\mu_e + \mu_h)$ and n_i is exponential

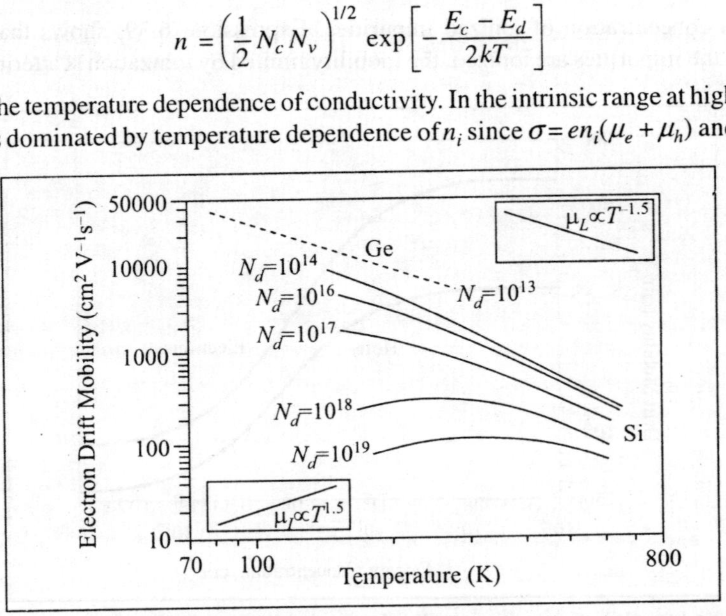

Fig. 6.17 Drift mobility variations according to doping concentration in silicon.

function of temperature in contrast to $\mu \propto T^{-1.5}$. In the extrinsic range $n = N_d$ and is constant, so that the conductivity follows the temperature dependence of the drift mobility.

6.5.3 Diffusion in Semiconductor

Diffusion is a process in which particles tend to spread out, and redistribute themselves as a result of their random thermal vibrations, migrating from region of high concentration to the region of low concentration. This process, if allowed to continue, will end only when there is uniform distribution of particles (same concentration) throughout the material. For diffusion to take place, the particle need not be charged, as thermal motion is responsible for this.

We will examine what happens to the particles confined in compartments shown in figure 6.18. Only the far left compartment has particle in it. The left wall of the compartment is so high that these particles cannot jump out of the compartment in that direction (left wall of the far left compartment). If τ is the mean free time of the particles, there is 50% probability that these particles will jump out of the compartment in either direction. Only in right hand side compartment, there will be half of the particle after τ seconds. After 2τ seconds, again the same process will repeat and there would be particles in 3rd compartment from the left. This process continues till there are equal particles in the entire compartments as the particles cannot jump out of the far right hand side wall as well.

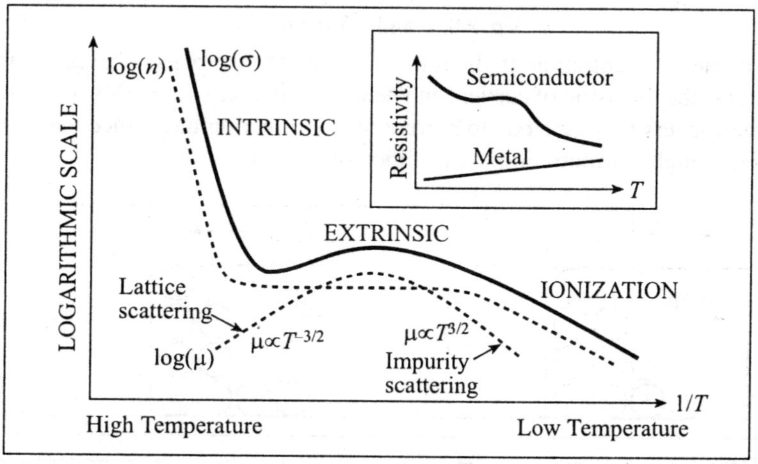

Fig. 6.18 Temperature dependence of conductivity in *a* n-type semiconductor.

In semiconductors, there is abundance of one of the charge carriers in doped semiconductors (fully uncompensated semiconductors). So, we can visualize a macroscopic diffusion from the region of high concentration of charge carriers to the region where their concentration is low In semiconductor, the particles are charged, which means their diffusion will give rise to electrical current called the diffusion current in semiconductor.

For diffusion to occur in semiconductor and give rise to diffusion current, there should be abundance of charge carriers in one region and scarcity in the other region. This means, there must be a concentration gradient along the axis along which diffusion is to take place.

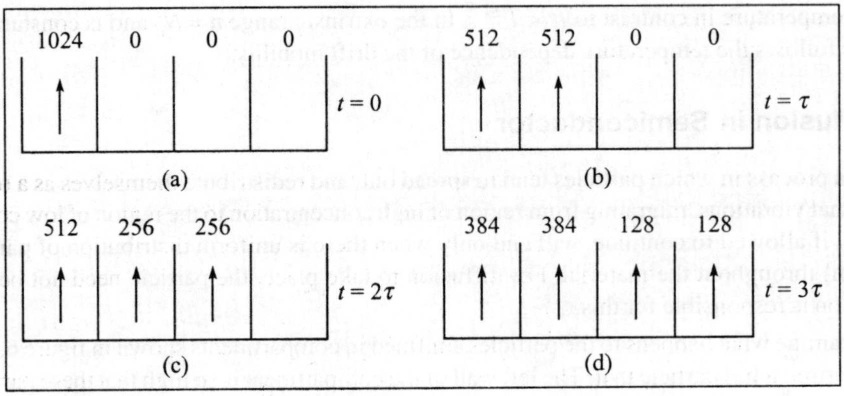

Fig. 6.19 Diffusion of particles in 1-dimensional compartments.

In single-dimension, for diffusion to take place the following condition must be met

$$\frac{dn}{dx} \neq 0, \text{ and } \frac{dp}{dx} \neq 0$$

And, for three-dimension, for diffusion to take place

$$\nabla p \neq 0, \quad \text{and} \quad \nabla n \neq 0$$

There must be carrier concentration gradient for diffusion to take place. The larger the concentration gradient, the greater is the diffusion of carriers and hence the greater will be flow of current. Diffusion of either of the charge carriers will give rise to current in one direction only since the carriers will always diffuse from region of high concentration to the region of low concentration.

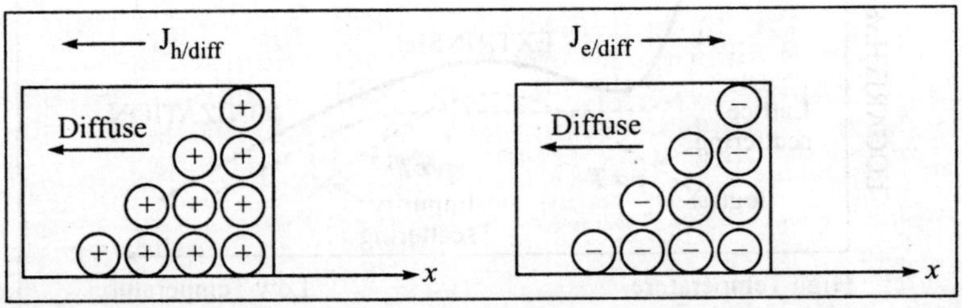

Fig. 6.20 Diffusion of charge carriers giving diffusion currents in doped semiconductors.

The diffusion currents are directly proportional to the carrier concentration gradients as given by

$$J_{e/diff} = eD_e \frac{dn}{dx} \tag{6.40}$$

$$J_{h/diff} = eD_h \frac{dp}{dx} \tag{6.41}$$

is also called Fick's first law of diffusion in one dimension, where D_e and D_h are electron and hole diffusion coefficients respectively measured in cm^2/sec. The negative sign in the expression for diffusion current density complies with the fact that for positive concentration gradient in x-direction (i.e., $dn/dx > 0$ and $dp/dx > 0$), the diffusion will take place in negative x-direction for both the holes and electrons. So, hole diffusion current will flow in negative x-direction (in the direction of diffusion) and electron diffusion current will flow in positive x-direction (in the direction opposite to the diffusion of electron).

For most materials, the diffusion coefficient is found to vary with temperature in an activated manner. Thus $D = D_0\, e^{-\phi/kT}$, where Φ is the activation energy and D_0 is pre-factor.

The total current density in the semiconductor due to electrons (both by drift and diffusion) is given by

$$J_e = J_{e/drift} + J_{e/diff} = en\mu_e E + eD_e \frac{dn}{dx} \tag{6.42}$$

Similarly, total current density in the semiconductor due to holes is given by

$$J_h = J_{h/drift} + J_{h/diff} = ep\mu_e E - eD_h \frac{dp}{dx} \tag{6.43}$$

Now, total current in semiconductor, which is the algebraic sum of contributions by both the electrons and holes, is given by

$$J = J_{drift} + J_{diff} = (J_{e/drift} + J_{h/drift}) + J_{e/diff} + J_{h/diff}) \tag{6.44}$$

$$J = (J_{e/drift} + J_{e/diff}) + (J_{h/drift} + J_{h/diff}) = J_e + J_h \tag{6.45}$$

6.5.4 Degenerate and Non-degenerate Semiconductors

The general exponential expression for the concentration of electron in conduction band

$$n \approx N_c \exp\left[-\frac{E_c - E_F}{kT}\right] \tag{6.46}$$

is based on replacing Fermi-Dirac statistics with Boltzmann statistics, which is only valid when E_c is several kT above E_F. We assumed that the number of states in the conduction band far exceeds the number of electrons there, so likelihood of two electrons trying to occupy the same state is almost none. So we can neglect Pauli's exclusion principle and can apply Boltzmann statistics for the electron concentration. N_c is the density of states in conduction band. So, the Boltzmann statistics is valid up to $n << N_c$. Semiconductors with $n << N_c$ and $p << N_v$ are termed non-degenerate semiconductors.

When the semiconductors have been excessively doped with, say, donors, electron concentration may reach a very large value of the order of $10^{19} - 10^{20}$ per cm^3. This electron concentration is comparable to density of states N_c in the conduction band edge. In this case, Pauli's exclusion principle can no longer be neglected and Fermi-Dirac statistics have to be used. Such semiconductors exhibit properties, which are more metal-like than semiconductor-like. For example, resistivity is now directly proportional to temperature. So, semiconductors with $n >> N_c$ and $p >> N_v$ are called degenerate semiconductors. In n-type semiconductor, when donor concentration is sufficiently high, the donor atoms become so close to each other that their orbitals overlap to form a narrow energy band that overlaps and becomes part of the conduction band thereby slightly shifting E_c downward. The valence electrons from the donor fill the conduction band right from E_c. The Fermi level in n-type degenerate semiconductor is within the conduction band. In other words, the Fermi level in degenerate n-type semiconductor lies above E_c and in degenerate p-type semiconductor it is below E_v.

6.6 Einstein Relationship

In a non-degenerate semiconductor, the Boltzmann statistics is obeyed and the Ferrmi level is at least ~ $3kT$ away from either conduction band-edge or valence band-edge. In such semiconductors, as doping level changes along the semiconductor, there will be corresponding change in the Fermi energy level.

For a non-degenerate semiconductor, the doping profile along its length is given by figure 6.21.

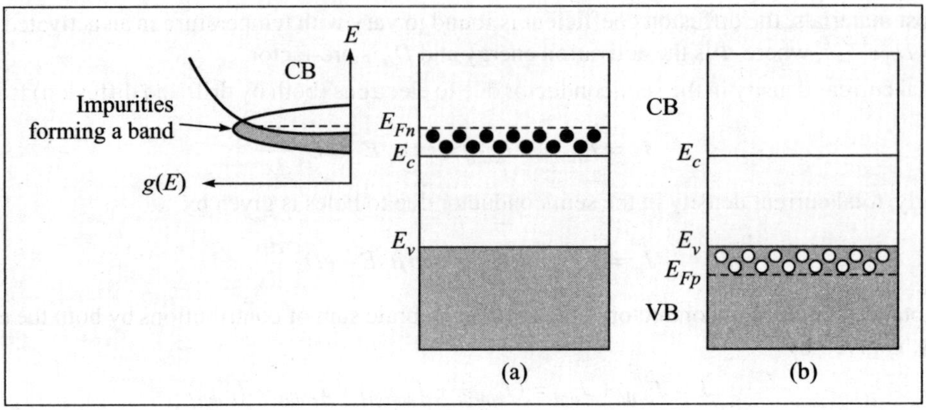

Fig. 6.21 Degenerate n-type semiconductor (a), and p-type semiconductor (b).

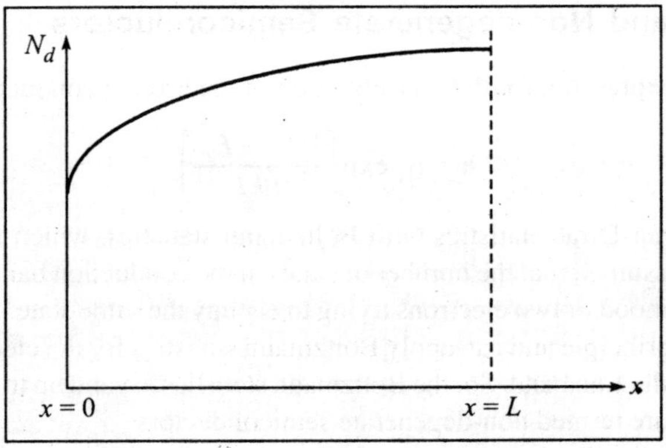

Fig. 6.22 Doping profile of an n-type non-degenerate semiconductor

But the energy band diagram shown in figure (6.23) is not accurate because it shows the variation in Fermi energy level within the semiconductor along its length. In equilibrium condition, Fermi level inside a material or in two materials in contact cannot vary with respect to position. Fermi energy level inside a material must be same. Mathematically

$$\frac{dE_F}{dx} = \frac{dE_F}{dy} = \frac{dE_F}{dz} = 0 \tag{6.47}$$

To keep the Fermi level same throughout the material, other energy bands- conduction band, valence band and intrinsic Fermi level have to bend, which is commonly referred as band bending in non-uniformly doped semiconductors.

Under equilibrium condition, the current in the semiconductor must be zero. The current due to drift and diffusion of electrons is given by equation 6.45. Recalling this equation, we have

$$J_e = (J_{eldrift} + J_{eldrift})$$

$$en\mu_e E + eD_e \frac{dn}{dx} = 0 \tag{6.48}$$

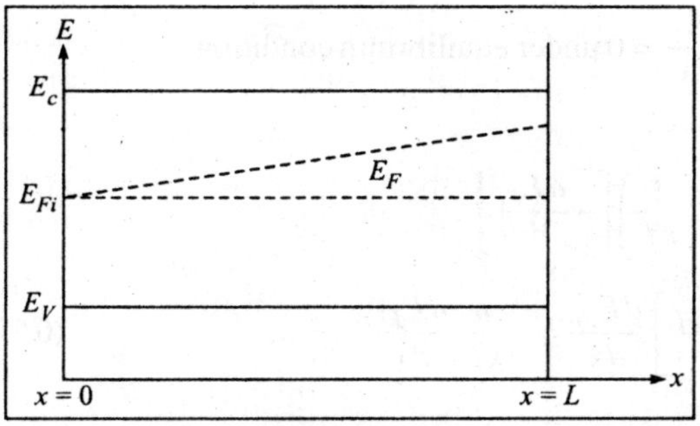

Fig. 6.23 Energy band diagram corresponding to doping profile in figure 6.22.

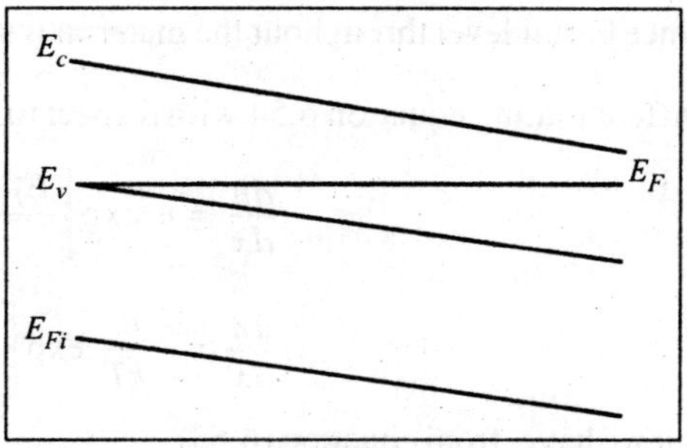

Fig. 6.24 Band bending in non-uniformly donor doped semiconductor

But the potential energy in electron-volt is related to electrostatic potential V at a point by

$$PE = -eV \quad (6.49)$$

Potential energy can also be expressed with respect to some reference as

$$PE = E_c - E_{ref} \quad (6.50)$$

From 6.49 and 6.50, we have

$$V = -\frac{1}{e}(E_c - E_{ref}) \quad (6.51)$$

The electric field is related to electrostatic potential as

$$\vec{E} = -\nabla V \text{ in 3-dimension}$$

$$\vec{E} = -\frac{dV}{dx} \text{ in 1-dimension} \quad (6.52)$$

From equation 6.51 and 6.52 for 1-dimension, we have

$$\vec{E} = \frac{1}{e}\frac{dE_c}{dx} = \frac{1}{e}\frac{dE_v}{dx} = \frac{1}{e}\frac{dE_{Fi}}{dx} \quad (6.53)$$

The electron concentration in n-type semiconductor is given by

$$n = N_c \exp\left[-\frac{E_c - E_F}{kT}\right]$$

and, the intrinsic concentration is

$$n_i = N_c \exp\left[-\frac{E_c - E_{Fi}}{kT}\right]$$

So

$$\frac{n}{n_i} = \frac{\exp\left[-\frac{E_c - E_F}{kT}\right]}{\exp\left[-\frac{E_c - E_{Fi}}{kT}\right]} = \exp\left[\frac{E_F - E_{Fi}}{kT}\right]$$

or,

$$n = n_i \exp\left[\frac{E_F - E_{Fi}}{kT}\right] \quad (6.54)$$

Since Fermi level throughout the material is same, $\frac{dE_F}{dx} = 0$ under equilibrium condition.

Differentiating equation 6.54 with respect to x, we have

$$\frac{dn}{dx} = n_i \exp\left[\frac{E_F - E_{Fi}}{kT}\right]\left[\frac{1}{kT}\right]\left[-\frac{dE_{Fi}}{dx}\right]$$

$$\frac{dn}{dx} = -\frac{n_i}{kT} \exp\left[\frac{E_F - E_{Fi}}{kT}\right]\frac{dE_{Fi}}{dx} = -\frac{n}{kT}\frac{dE_{Fi}}{dx} \tag{6.55}$$

But we have, from equation (6.53)

$$\frac{1}{e}\frac{dE_{Fi}}{dx} = \vec{E}$$

or,

$$\frac{dE_{Fi}}{dx} = e\vec{E}$$

Putting this in equation (6.55) yields

$$\frac{dn}{dx} = -\frac{n}{kT}\, e\vec{E} \tag{6.56}$$

Substituting from equation (6.56) in equation 6.48, we get

$$en\mu_e\vec{E} + eD_e\left(-\frac{n}{kT}e\vec{E}\right) = 0$$

or,

$$en\left(\mu_e - \frac{D_e}{kT}e\right)\vec{E} = 0$$

or,

$$\mu_e = \frac{D_e}{kT}e$$

or,

$$\frac{kT}{e} = \frac{D_e}{\mu_e} \tag{6.57}$$

Equation (6.57) relating electron diffusion coefficient in n-type semiconductor to electron mobility is called Einstein relationship. For a given n-type semiconductor material at a constant temperature, the ratio of diffusion coefficient to mobility of electron is constant and given by kT/e.

Similarly, for p-type semiconductor, diffusion coefficient and mobility of hole are related to each other in the following manner

$$\frac{kT}{e} = \frac{D_h}{\mu_h} \tag{6.58}$$

The equation (6.57) derived above and known as Einstein's equation can also be derived in the following manner:

Recalling the expression (6.48), we have

$$en\mu_e E + eD_e\frac{dn}{dx} = 0$$

or,

$$en\mu_e E = -eD_e\frac{dn}{dx}$$

or,
$$\frac{dn}{n} = -\mu_e E dx$$

Integrating both sides, we get

$$\ln(n) = -\frac{\mu_e Ex}{D_e} + A$$

or,
$$n = A \exp\left(-\frac{\mu_e xE}{D_e}\right) \quad (6.59)$$

According to Boltzmann's statistics, for an electric field we have

$$n = A \exp\left(-\frac{exE}{kT}\right) \quad (6.60)$$

Hence from equations (5.59) and (5.60), we get

$$-\frac{\mu_e xE}{D_e} = \frac{exE}{kT}$$

i.e.
$$\frac{\mu_e}{D_e} = \frac{e}{kT} \quad (6.61)$$

The above equation is extensively used in ionic system.

6.7 Recombination and Generation

6.7.1 Direct Recombination and Generation

Above absolute zero temperature, the thermal excitation of electrons from the valence band to the conduction band continuously generates electron-hole pairs, i.e. conduction electron and valence hole pairs. In equilibrium condition, there is an annihilation mechanism by which conduction electron returns to valence hole and fills this empty state. So, when a conduction electron wandering freely in that band falls down to the low energy empty electronic state and meets the hole there, the process is called ***recombination***. By this process free electron in conduction band and free hole in valence band are annihilated or destroyed. In figure 6.25 direct recombination mechanism in GaAs is shown. The excess energy of electron is lost as a photon of energy $h\nu = E_g$. Such recombination process results in the emitted light from light emitting diodes (LEDs).

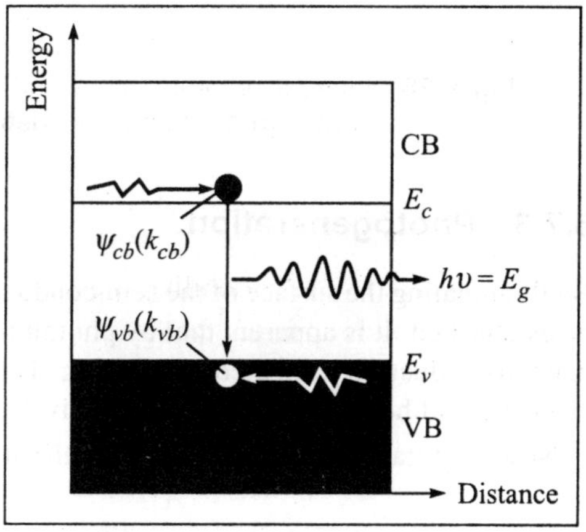

Fig. 6.25 Direct recombination in GaAs

6.7.2 Indirect Thermal Generation and Recombination

In elemental semiconductor crystals like Si and Ge, electrons and holes usually recombine through recombination centers. A recombination center can take up any momentum difference between a hole and electron, so increases the probability of recombination. This process essentially involves a third party, which may be an impurity atom or crystal defect. The electron is captured at the recombination center and localized at this site. The electron is held or trapped at this center until a hole arrives and recombines with it. The recombination center provides localized electronic state below E_c in the band gap. In figure 6.26 recombination center is shown. When an electron approaches the center, it is captured, and is now bound to the local state. The localized electron waits the arrival of hole and recombines during which some energy is lost lattice vibrations via third party. Emitted lattice vibrations are called ***phonons***. A phonon is a quantum energy associated with atomic vibration in crystal analogous to photon. Typical recombination centers might be metallic impurities and crystal defects.

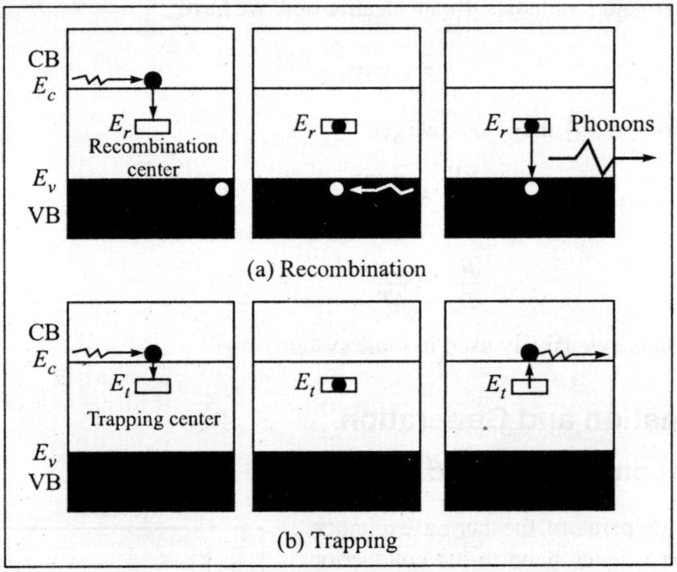

Fig. 6.26 Indirect recombination and generation. Recombination in Si via recombination centre (a);trapping and detrapping of electrons by trapping centres (b).

6.7.3 Photogeneration

By illuminating the surface of the semiconductor, a steady state non-equilibrium carrier concentration can be established. It is apparent that by photon of sufficient energy we can promote electrons from valence band to conduction band thereby creating electron-hole pair. Let the thermal equilibrium concentration of electrons and holes be n_0 and p_0 respectively, and the excess concentrations are Δn and Δp.

Now the total electron and hole concentrations are

$$n = n_0 + \Delta n \tag{6.62}$$

$$p = p_0 + \Delta p \tag{6.63}$$

Photogeneration creates equal number of electron and holes.

$$\Delta n = \Delta p$$

The excess concentrations for example reached at a time $t = 0$, can be denoted by $\Delta n(0)$ and $\Delta p(0)$ are equal to each other and are the values of excess concentrations at $t = 0$ for steady state condition.

If steady state has been reached before $t = 0$ and the source of illumination is removed at $t = 0$, then the excess carrier concentration will decrease to an equilibrium value with a time constant τ_h called excess carrier lifetime or minority carrier lifetime or mean recombination time. This is the average time for which a hole exists in valence band from its generation to recombination, that is, the mean time the hole is free before recombining with an electron. Although the excess majority charge carriers also die along with excess minority carrier, the time during which excess carriers die is referred as minority carrier lifetime because there is still very large number of majority carriers whereas the number of minority carriers is drastically reduced

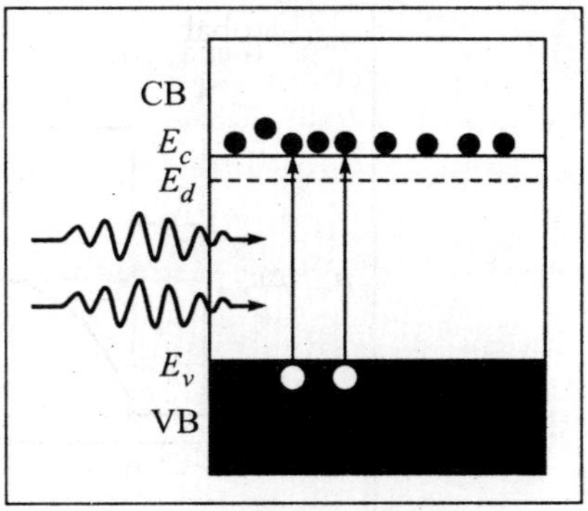

Fig. 6.27 Low level photo-injection into an n-type semiconductor where D$n <$ n_0

Rate of increase in hole concentration is equal to the difference of rate of photogeneration and rate of recombination of excess holes with electrons.

$$\frac{d\Delta p}{dt} = G_{ph} - \frac{\Delta p}{\tau_h} \tag{6.64}$$

where G_{ph} is the rate of photogeneration of holes,

$\Delta p/\tau_h$ represents rate of recombination of excess holes.

The recombination time τ_h depends on semiconductor material, impurities, crystal defects, temperature and so forth, and therefore can not be quoted by a typical value but is somewhere from nanoseconds to seconds.

Consider an n-type semiconductor where illumination is switched on at $t = 0$ and switched off at $t = t_{off.}$ Usiing the fact that G_{ph} = const in $0 < t < t_{off}$, integrating equation (6.64), we have

$$\ln [G_{ph}\, \tau_h - \Delta p] = -\frac{t}{\tau_h} + C$$

where C is the constant of integration. At $t = 0$, $\Delta p = 0$, so that $C = G_{ph.}$

The solution is

$$\Delta p(t) = \tau_h\, G_{ph}\, [1 - \exp(-t/\tau_h)] \text{ in the region } 0 < t < t_{off}.$$

As soon as the illumination is switched on, the minority carrier concentration rises exponentially toward its steady-state value $\Delta p(\infty) = \tau_h\, G_{ph}$ which is reached after $t > \tau_h$.

When illumination is switched off at $t_{off} >> \tau_h$, $\Delta p(t_{off}) = \tau_h\, G_{ph}$

Let t' be the time measured from $t = t_{off}$, then $t' = t - t_{off}$.

$$\Delta p(t' = 0) = \tau_h G_{ph}$$

Solving equation 6.61 with $G_{ph} = 0$ in $t > t_{off}$ or $t' > 0$, we have

$$\Delta p(t') = \Delta p(0) \exp(-t'/\tau_h) \tag{6.65}$$

Differentiating equation (6.65) with respect to t, we get

$$\frac{d\Delta p(t')}{dt} = -\frac{1}{\tau_h} \Delta p(0) \exp\left[-\frac{t}{\tau_h}\right]$$

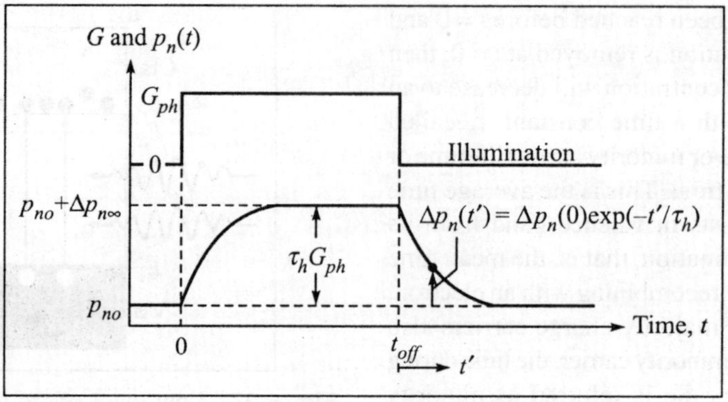

Fig. 6.28 Hole concentration profile when illumination is switched on at $t = 0$ and switched off at $t = t_{off}$. Excess minority carrier concentration (of holes) both rises and decays exponentially.

or,
$$\frac{d\Delta p}{dt} = \frac{\Delta p(t')}{\tau_h} \tag{6.66}$$

which also gives the rate of recombination of excess minority carriers.

6.7.4 Continuity Equations for Carriers

As shown in the figure the hole current density entering the sample $J_h(x)$ can be greater or smaller than the current density leaving the sample $J_h(x + \Delta x)$ depending on the generation or recombination taking place within the sample volume. The net increase in hole concentration per unit time is the difference between the hole flow per unit volume entering and leaving minus the rate of recombination as given by equation (6.64).

The hole current density entering the area A can be converted to hole flow by diving by electron charge, i.e. $J_h(x)/e$. And, by dividing hole flow entering the area A by Δx, we will get number of carriers per unit volume entering the sample $J_h(x)/(e\Delta x)$. Similarly, the hole flow leaving the volume per unit time is $J_h(x + \Delta x)/e$ and number of holes per unit volume per unit time leaving the sample is $J_h(x + \Delta x)/(e\Delta x)$.

$$\left.\frac{\partial \Delta p}{\partial t}\right|_{x\to\Delta x} = \frac{1}{e}\,\frac{J_h(x) - J_h(x+\Delta x)}{\Delta x} - \frac{\Delta p}{\tau_h} \tag{6.67}$$

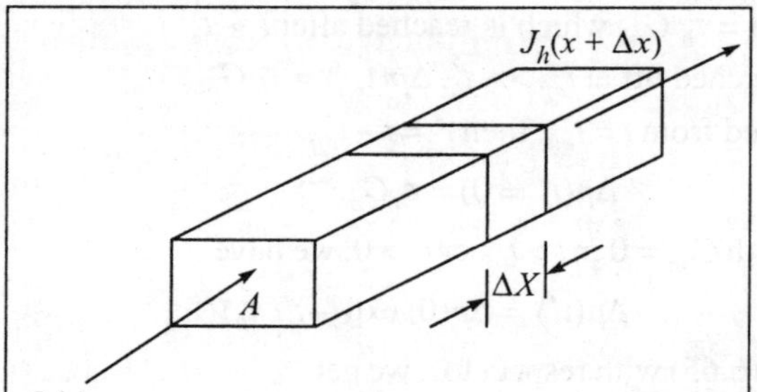

Fig. 6.29 Particle current entering and leaving the sample.

The term in left hand side gives the rate of hole build-up where as the first term in right hand side represents increase in hole concentration in volume $A\Delta x$ per unit time, and the last term is recombination rate.

But, limit $\Delta x \rightarrow 0$

$$\frac{\partial p(x,t)}{\partial t} = -\frac{1}{e}\frac{\partial J_h}{\partial x} - \frac{\Delta p}{\tau_h} \tag{6.68}$$

The equilibrium concentration is independent of time. P_0 = constant. So,

$$\frac{\partial p}{\partial t} = \frac{\partial \Delta p}{\partial t} \tag{6.69}$$

Now from the above two expressions, we have

$$\frac{\partial \Delta p}{\partial t} = -\frac{1}{e}\frac{\partial J_h}{\partial x} - \frac{\Delta p}{\tau_h} \tag{6.70}$$

Equation (6.70) is called the ***continuity equation*** for ***holes***.

Similarly, for electrons we have

$$\frac{\partial n}{\partial t} = -\frac{\Delta n}{\tau_e} + \frac{1}{e}\frac{\partial J_e}{\partial x} \tag{6.71}$$

is the ***continuity equation*** for ***electrons***. The positive sign before the last term in right hand side of the expression signifies the opposite nature of flow of electron and resulting current density.

Recalling the expression for hole current density from equation (6.43), we have

$$J_h = J_{h/drift} + J_{h/diff} = ep\mu_e E - eD_h \frac{dp}{dx}$$

Now substituting for J_h in equation (6.67), we will have

$$\frac{\partial \Delta p}{\partial t} = -\frac{\Delta p}{\tau_h} - \frac{1}{e}\frac{\partial}{\partial x}\left(ep\mu_h \vec{E} - eD_h \frac{dp}{dx}\right) \tag{6.72}$$

In the above expression for uniformly doped semiconductor, $\vec{E}$ is small. So the term consisting $\vec{E}$ can be neglected from the expression. So,

$$\frac{\partial \Delta p}{\partial t} = -\frac{\Delta p}{\tau_h} + D_h \frac{\partial^2 \Delta p}{\partial x^2} \tag{6.73}$$

Equation (6.73) is called Fick's 2nd law of diffusion (or ***diffusion equation*** for ***holes***).

For metals, $\frac{\Delta p}{\tau_h}$ and $p = n$

Under steady-state condition, i.e. $\frac{\partial \Delta p}{\partial t} = 0$, equation (6.73) becomes

$$\frac{d^2 \Delta p}{dx^2} - \frac{\Delta p}{\tau_h D_h} = 0 \tag{6.74}$$

$$\frac{d^2 \Delta p}{dx^2} - \frac{\Delta p}{L_h^2} = 0 \tag{6.75}$$

where $L_h = \sqrt{\tau_h D_h}$ is called diffusion length for the holes in n-type semiconductor. L_h is the length a hole diffuses in average before it recombines with an electron in n-type semiconductor, that is why, L_h is also called the minority carrier diffusion length.

Solution of the equation (6.75) is in the form

$$\Delta p = A \exp\left(\frac{x}{L_h}\right) + B \exp\left(-\frac{x}{L_h}\right) \tag{6.76}$$

Similarly, for electrons in p-type semiconductor, we can write second order differential equation in the form

$$\frac{d^2 \Delta n}{dx^2} - \frac{\Delta n}{L_e^2} = 0 \tag{6.77}$$

where $L_e = (\tau_e D_e)^{1/2}$ is called diffusion length of electrons in p-type semiconductor. The solution of expression (6.77) is

$$\Delta n = A \exp\left(\frac{x}{L_e}\right) + B \exp\left(-\frac{x}{L_e}\right) \tag{6.78}$$

The diffusion length of injected carriers is important in determining certain dimensions in semiconductor devices such as diodes, transistors, etc. For example, for efficient operation of transistor, the width of base region must be small compared to the diffusion length of injected carriers. Otherwise a large portion of them would be lost by recombination before reaching the collector-base junction.

6.8 PN Junction

6.8.1 Ideal Abrupt Junction

In an n-type silicon semiconductor electrons are majority charge carriers and holes are minority charge carriers. Similarly, in a p-type semiconductor, holes are majority carriers and electrons are minority carriers.

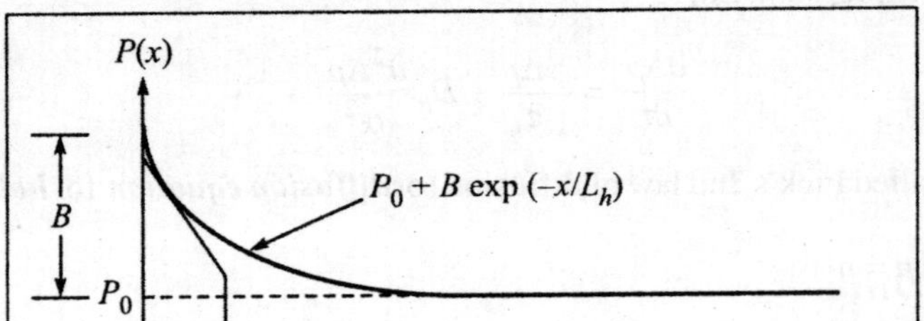

Fig. 6.30 Steady-state minority carrier concentration profile in an n-type semiconductor illuminated at one end.

Let us see what happens when these two semiconductors are brought together to form an abrupt junction (there is abrupt discontinuity between p and n regions). The abrupt junction is called metallurgical junction.

As shown in figure 6.31.a an n-type and a p-type semiconductor are brought together. The n-region has excess electrons in conduction band and immobile positive ions whereas the p-region has excess free holes and immobile negative ions. The electron concentration and hole concentrations in n-region are n_{n0} and p_{n0} respectively. Similarly, the hole and electron concentrations in p-region are p_{p0} and n_{p0} respectively. Due to concentration gradient of holes from p region to n region, they will diffuse toward n-region and electrons will diffuse toward p region thanks to the concentration gradient from n region to p region. These holes and electrons diffusing toward each other meet and recombine around the junction. The junction region consequently becomes depleted of free charge carriers in comparison to the bulk n and p regions away from the junction. The region around junction is thus depleted of free charge carriers, so it is called depletion layer. The free charge carriers diffusing to p and n regions leave behind exposed positively charged impurity ions at or around the junction. So, this region is also called space charge region or layer

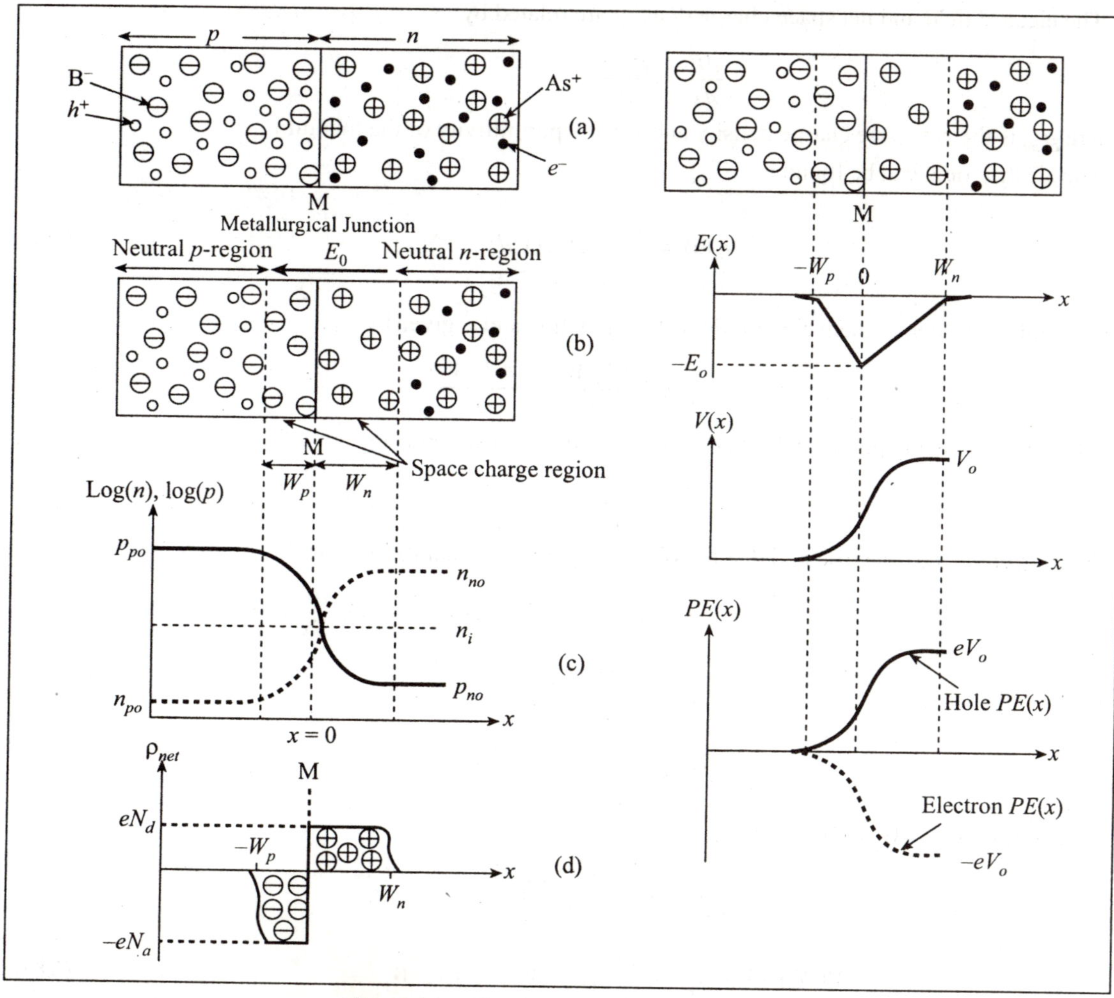

Fig. 6.31 Abrupt idealized *pn* junction

(SCL). Through this region, the concentration of electron and hole changes in such a way that they transform from majority to minority carriers and vice versa. So, this region is also called transition region. Figure 6.31.b shows the space charge region and figure 6.31.d shows the carrier concentration profile in logarithmic scale. Here the acceptor concentration is greater than the donor concentration.

Even without applied field, there is an internal field E_0 directed from positive ions to negative ions. This internal field tries to drift holes back to the p-region and electrons back to the n-region. In other words, the internal field tries to drive the carriers in direction opposite to their diffusion. The strength of the field will be determined by the equilibrium condition at which the number of carriers diffusing is compensated by the number of carriers drifting in opposite direction. For uniformly doped p and n regions, the net space charge density is as shown in figure 6.31.d The exposed charged in the left side of the junction must be equal to the exposed charge in the right side of the junction to maintain the charge neutrality.

$$N_a W_p = N_d W_n \tag{6.79}$$

Since $N_a > N_d$, $W_p < W_n$.

The electric field and net space charge density are related by

$$\frac{dE}{dx} = \frac{\rho_{net}}{\varepsilon} \tag{6.80}$$

where ρ_{net} is the net space charge density and ε is the permittivity of the medium.

The electric field can be found as

$$\vec{E}(x) = \frac{1}{\varepsilon}\int_{-W_p}^{x} \rho_{net}(x)dx \tag{6.81}$$

which will give the maximum electric field at the junction and is given by

$$\vec{E}_0 = -\frac{eN_d W_n}{\varepsilon} = \frac{eN_a W_d}{\varepsilon} \tag{6.82}$$

The potential at any point can be calculated from the electric field.

$$\vec{E} = -\frac{dV}{dx}$$

Net current due to electron and due to hole are separately equal to zero. So,

$$J_e = J_{eldrift} + J_{eldiff} = en\mu_e \vec{E} + eD_e \frac{dn}{dx} = 0$$

$$\vec{E} = -\frac{D_e}{n\mu_e}\frac{dn}{dx}$$

$$-\frac{dV}{dx} = -\frac{kT}{en}\frac{dn}{dx}$$

Integrating from $-W_p$ to W_n, we have

$$\int_{-W_p}^{W_n} dV = \frac{kT}{e}\int_{-W_p}^{W_n}\frac{dn}{n}$$

or,

$$V(W_n) - V(-W_p) = \frac{kT}{e}[\ln n(W_n) - \ln n(-W_p)) \tag{6.83}$$

$$V_0 = V(W_n) - V(W_p) = \text{Built-in potential}$$

The electron concentration at $x = -W_p$ is $n_{p0} = \frac{n_i^2}{N_a}$ and at $x = W_n$, $n_{n0} = N_d$

Putting these values in equation (6.80), we will find the built-in potential.

$$V_0 = \frac{kT}{e}\left[\ln N_d - \ln \frac{n_i^2}{N_d}\right]$$

or,

$$V_0 = \frac{kT}{e} \ln \frac{N_a N_d}{n_i^2} \tag{6.84}$$

At $x = -W_p$, $n_{p0} = n_i^2/N_a$ and at $x = W_n$, $n_{n0} = N_d$. Putting these values in equation (6.84) yields

$$V_0 = \frac{kT}{e} \ln\left(\frac{n_{n0}}{n_{p0}}\right) \tag{6.85}$$

From which we can find

$$\frac{n_{n0}}{n_{p0}} = \exp\left(\frac{eV_0}{kT}\right) \tag{6.86}$$

Similarly we can write

$$V_0 = \frac{kT}{e} \ln\left(\frac{p_{p0}}{p_{n0}}\right) \tag{6.87}$$

And

$$\frac{p_{p0}}{p_{n0}} = \exp\left(\frac{eV_0}{kT}\right) \tag{6.88}$$

In the above expressions, n_{n0} and n_{p0} represent electron concentrations in bulk of n-region and p-region respectively. Similarly, p_{p0} and p_{n0} represent hole concentration in bulk p and n-region respectively.

The built in potential

$$V_0 = -\frac{\vec{E}_0 W_0}{2} = \frac{eN_d W_n}{2\varepsilon} W_0 \tag{6.89}$$

Where $W_0 = W_p + W_n$ is the depletion width. So, $W_p = W_0 - W_n$

We also have,

$$W_n N_d = W_p N_a$$

$$W_n N_d = (W_0 - W_n) N_a = W_0 N_a - W_n N_a$$

or,

$$W_n = \frac{W_0 N_a}{N_a + N_a} \tag{6.90}$$

Now substituting W_n from equation (6.90) in to equation (6.89), we have

$$V_0 = \frac{eN_a N_d W_0^2}{2\varepsilon(N_a + N_d)} \tag{6.91}$$

Depletion width can be expressed as

$$W_0 = \left\{\frac{2\varepsilon V_0}{e} \frac{(N_a + N_d)}{N_a N_d}\right\}^{1/2} = \sqrt{\left\{\frac{2\varepsilon V_0}{e}\left(\frac{1}{N_a} + \frac{1}{N_d}\right)\right\}} \tag{6.92}$$

Substituting for V_0 from equation (6.84) in to equation (6.92), we have

$$W_0 = \sqrt{\frac{2\varepsilon kT}{e^2}\ln\frac{N_a N_d}{n_i^2}\left(\frac{1}{N_a}+\frac{1}{N_d}\right)} \tag{6.93}$$

If one of the impurity concentrations is very small compared to the other, let $N_a << N_d$, then built-in potential and depletion width can be found as

$$V_0 = \approx \frac{eN_a W_0^2}{2\varepsilon} \tag{6.94}$$

$$W_0 = \sqrt{\frac{2\varepsilon kT}{e^2}\ln\frac{N_d}{n_i^2}} \tag{6.95}$$

In this case, almost all of the depletion region will fall in p-region as doping concentration of n-region is much heavier than that of p-region, which requires charge from very small range in n-region to be balanced by charge from relatively larger range in p-region to balance the former.

6.8.2 Forward Biased pn Junction

When there is no external voltage applied to the ***pn*** junction, there will be no current through it. When an external voltage source (battery) is connected to the ***pn*** junction in such a manner that the positive terminal of the battery is connected to far-end of p-region and negative terminal is connected to far end of *n*-region, then the ***pn*** junction is called forward biased. The internal electric field and built-in potential both will be reduced as shown in figure 6.32. The new electric field is still directed from right of the junction (depletion region in n-side) to left of the junction (to the depletion region in p-side) because $E_0 > E$. The applied voltage will mostly drop in the depletion region as it has high resistivity compared to the bulk because there is far higher number of free charge carriers in bulk compared to immobile ions in the depletion region. So, applied voltage directly opposes built-in potential. Consequently, built-in potential is reduced from V_0

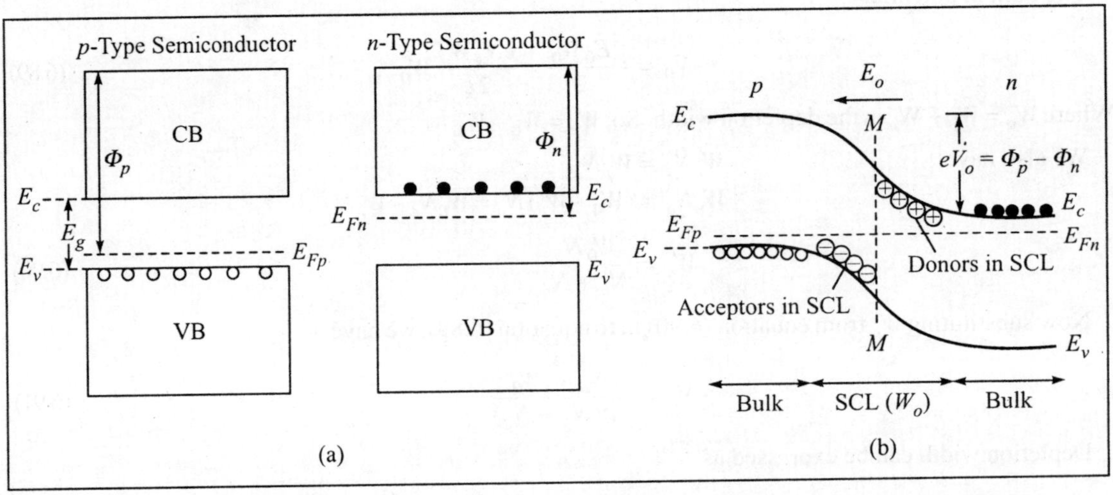

Fig. 6.32 Energy band diagram of two isolated p and n-type semiconductors (a). Energy band diagram when p and n-type semiconductors are in contact (b).

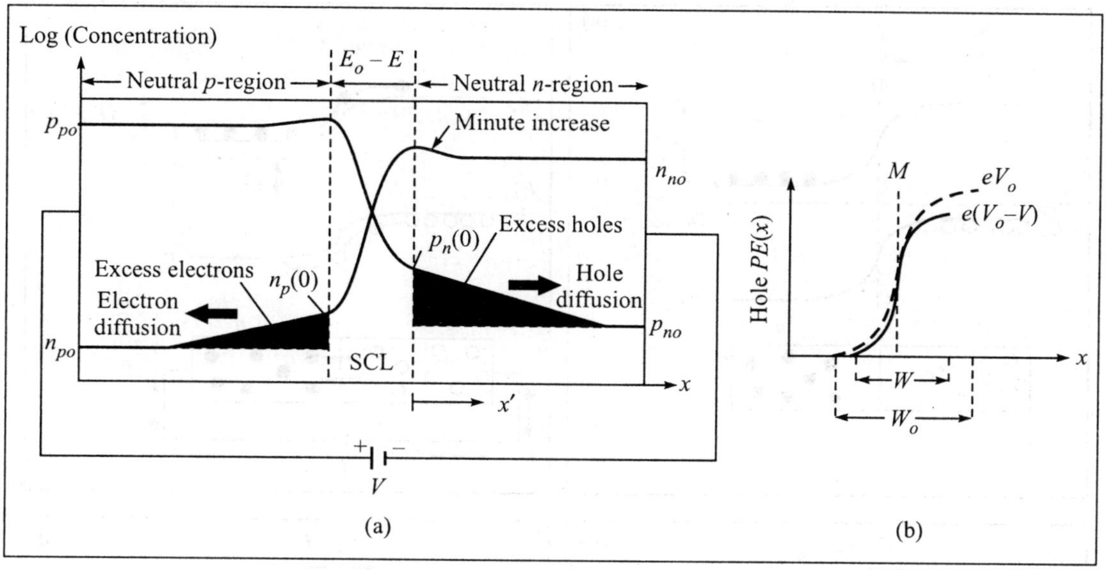

Fig. 6.33 Forward biased *pn* junction. Carrier concentration profile across the device (a); hole potential energy with or without bias.

to new value $(V_0 - V)$. The potential barrier to diffusion is then reduced which has drastic consequences. The probability that a hole will surmount this height and diffuse to n-region is now proportional to exp $[-e(V_0 - V)/kT]$. Many holes will diffuse across depletion region and enter the n-side. This is like injection of excess minority carriers, holes, into the n-region. In the similar manner electrons are injected into p-region.

The hole concentration in n-region is $p_n(0) = p_n(x' = 0)$ just outside the depletion region at $x' = 0$ is due to excess of holes diffusing as a result of reduction in- built-in potential barrier. This concentration $p_n(0)$ is found from probability of surmounting the potential energy barrier $e(V_0 - V)$ and given by

$$p_n(0) = p_{p0} \exp\left[-\frac{e(V_0 - V)}{kT}\right] \tag{6.96}$$

But, we have, recalling equation (6.88)

$$\frac{p_{p0}}{p_{n0}} = \exp\left(\frac{eV_0}{kT}\right)$$

or,

$$p_{p0} = p_{n0} \exp\left(\frac{eV_0}{kT}\right)$$

Substituting this in equation (6.96), we get

$$p_n(0) = p_{n0} \exp\left(\frac{eV_0}{k}\right) \exp\left[-\frac{e(V_0 - V)}{kT}\right] = p_{n0} \exp\left(\frac{eV}{kT}\right) \tag{6.97}$$

The expression (6.97) represents what is called "Law of junction".

Injected holes diffuse in n-region and eventually recombine with electrons in this region, as there are plenty of electrons there. These electrons lost by recombination with holes are readily replenished by negative terminal of the battery connected to n-side. As the p-region can supply more and more holes to

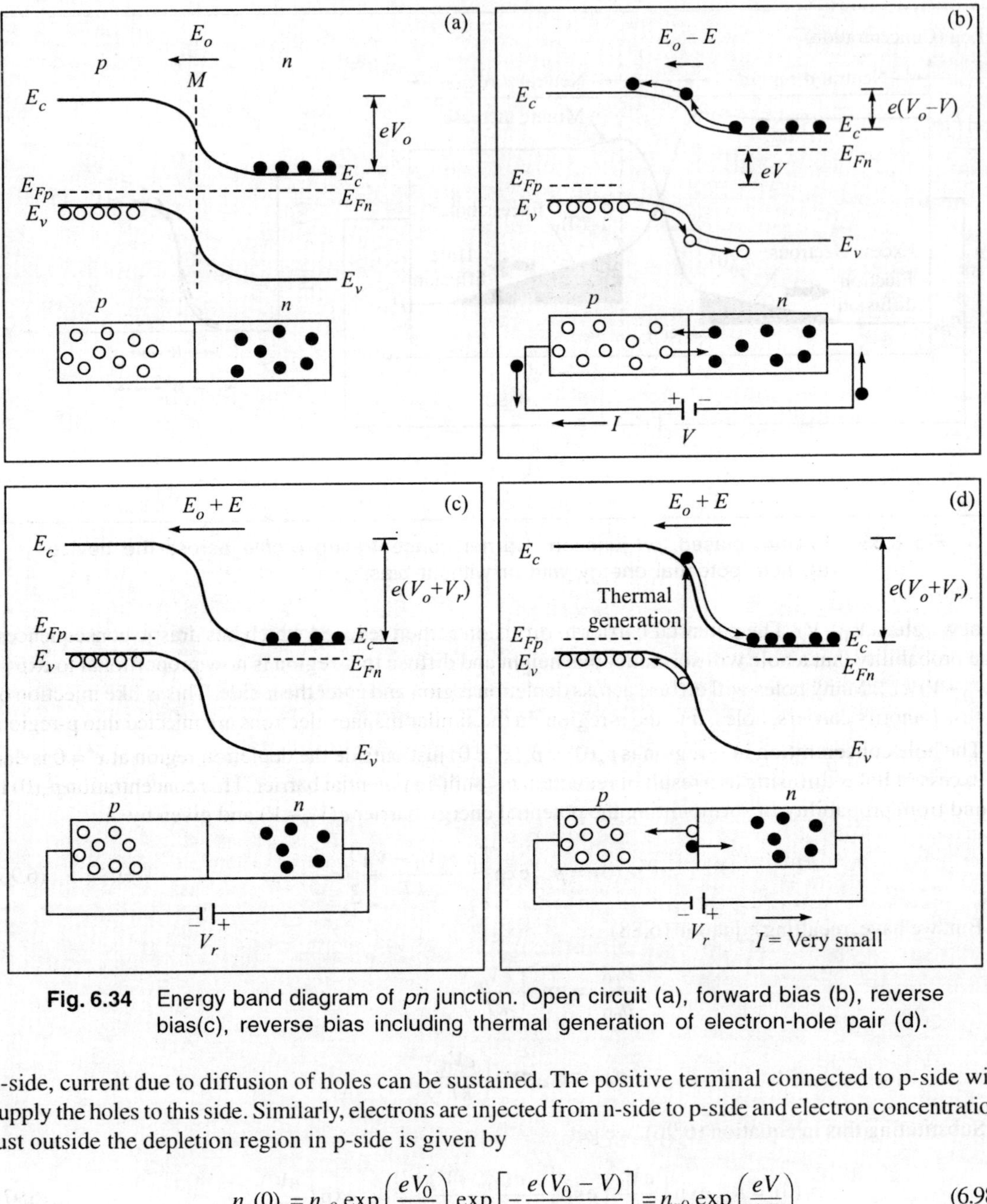

Fig. 6.34 Energy band diagram of *pn* junction. Open circuit (a), forward bias (b), reverse bias(c), reverse bias including thermal generation of electron-hole pair (d).

n-side, current due to diffusion of holes can be sustained. The positive terminal connected to p-side will supply the holes to this side. Similarly, electrons are injected from n-side to p-side and electron concentration just outside the depletion region in p-side is given by

$$n_p(0) = n_{p0} \exp\left(\frac{eV_0}{kT}\right) \exp\left[-\frac{e(V_0 - V)}{kT}\right] = n_{p0} \exp\left(\frac{eV}{kT}\right) \tag{6.98}$$

By forward biasing a *pn* junction, a current through the junction can be maintained. Holes diffusion in n-side and electrons diffusing in p-side will give rise to current in the same direction. The current flow seems to be surprisingly due to the diffusion of minority charge carriers, there is, however, current due to drift of majority carriers as well.

If the length of p-side and n-side are longer than minority carrier diffusion lengths, then the hole concentration $p_n(x')$ on the n-side will fall exponentially toward thermal equilibrium value $p_{n0.}$ So,

$$\Delta p(x') = \Delta p_n(0)\exp\left(-\frac{x'}{L_h}\right) \tag{6.99}$$

where $\Delta p_n(x') = p_n(x') - p_{n0}$ is the excess carrier distribution and L_h is the hole diffusion length. The hole current density is

$$J_{h/Driff} = -eD_h\frac{dp_n(x')}{dx'} = -eD_h\frac{d\Delta p_n(x')}{dx'}$$

or,

$$J_{h/driff} = \frac{eD_h}{L_h}\Delta p_n(0)\exp\left(-\frac{x'}{L_h}\right) \tag{6.100}$$

At $x' = 0$ just outside the depletion region, the hole diffusion current is

$$J_{h/driff} = \frac{eD_h}{L_h}\Delta p_n(0)$$

But

$$\Delta p_n(0) = p_n(0) - p_{n0} = p_{n0}\exp\left(\frac{eV}{kT}\right) - p_{n0} = p_{n0}\left[\exp\left(\frac{eV}{kT}\right) - 1\right]$$

Putting this in to above equation will give

$$J_{h/driff} = \frac{eD_h}{L_h}p_{n0}\left[\exp\left(\frac{eV}{kT}\right) - 1\right] \tag{6.101}$$

But

$$p_{n0} = \frac{n_i^2}{n_{n0}} = \frac{n_i^2}{N_d}$$

$$J_{h/driff} = \frac{eD_h n_i^2}{L_h N_d}\left[\exp\left(\frac{eV}{kT}\right) - 1\right] \tag{6.102}$$

Similarly, the diffusion current density for electrons is given by

$$J_{eldiff} = \frac{eD_e n_i^2}{L_e N_a}\left[\exp\left(\frac{eV}{kT}\right) - 1\right] \tag{6.103}$$

The total diffusion current density is

$$J_{diff} = J_{eldiff} + J_{h/diff} = en_i^2\left(\frac{D_e}{L_e N_a} + \frac{D_h}{L_h N_d}\right)\left[\exp\left(\frac{eV}{kT}\right) - 1\right]$$

or,

$$J_{diff} = J_{s0}\left[\exp\left(\frac{eV}{kT}\right) - 1\right] \tag{6.104}$$

Equation (6.104) is called ***ideal diode equation***. In terms of current, this can be expressed as

$$I_{diff} = J_{diff}A = AJ_{s0}\left[\exp\left(\frac{eV}{kT}\right) - 1\right] = I_{s0}\left[\exp\left(\frac{eV}{kT}\right) - 1\right] \tag{6.105}$$

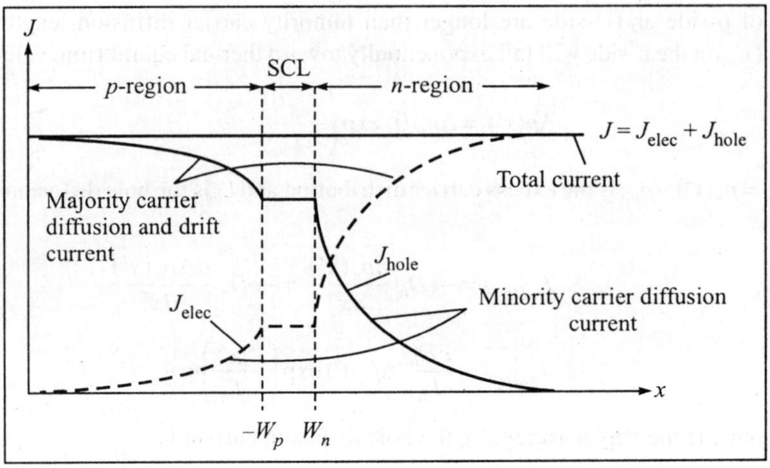

Fig. 6.35 Current through the *pn* junction during forward bias condition.

Where
$$J_{s0} = \left(\frac{D_e}{L_e N_a} + \frac{D_h}{L_h N_d}\right) e n_i^2 \tag{6.106}$$

J_{s0} is called reverse saturation current density. And I_{s0} is reverse saturation current.

6.8.3 Reverse Biased *pn* Junction

When pn junction is connected to a voltage source with p-region connected to negative terminal of voltage source and n-region to positive terminal, then the junction is said to be reverse biased.

The negative terminal of the battery will attract holes in p-region to move away from the depletion layer resulting in more exposed negative acceptor ions and thus a wider space charge layer. In the same way, the positive terminal will attract electrons away from the depletion region exposing more positively charged donors and consequently widening the depletion region in this side as well. The movement of electrons in the n-region toward the positive battery terminal cannot be sustained, as there is no electron supply to this side. And, p-side cannot supply electrons to n-side, as it has almost none. The electrons from the battery cannot enter the p-side and transit across to n-side because before they get a chance to do so, they will be recombined with holes. So, a current cannot be sustained through pn junction in reverse baised condition. The applied voltage adds to the built-in potential. Similarly, the electric field in the depletion layer is the sum of internal field and applied field. The holes in the n-side near the depletion layer can be extracted by the field and driven across depletion layer top the p-region. These holes can exit the p-region and enter the battery, giving rise to a very small current. Similarly, the electrons in p-side near the depletion region are extracted and driven away through the depletion layer onto n-side and again give rise to a small current. There will be small diffusion current through the junction due to holes diffusing from n-region to depletion layer and and electrons diffusing to depletion layer from p-region.

Recalling the expression (6.106) the reverse saturation current density is given by

$$J_{s0} = \left(\frac{D_e}{L_e N_a} + \frac{D_h}{L_h N_d}\right) e n_i^2 \tag{6.106}$$

is called ***Shockly equation***.

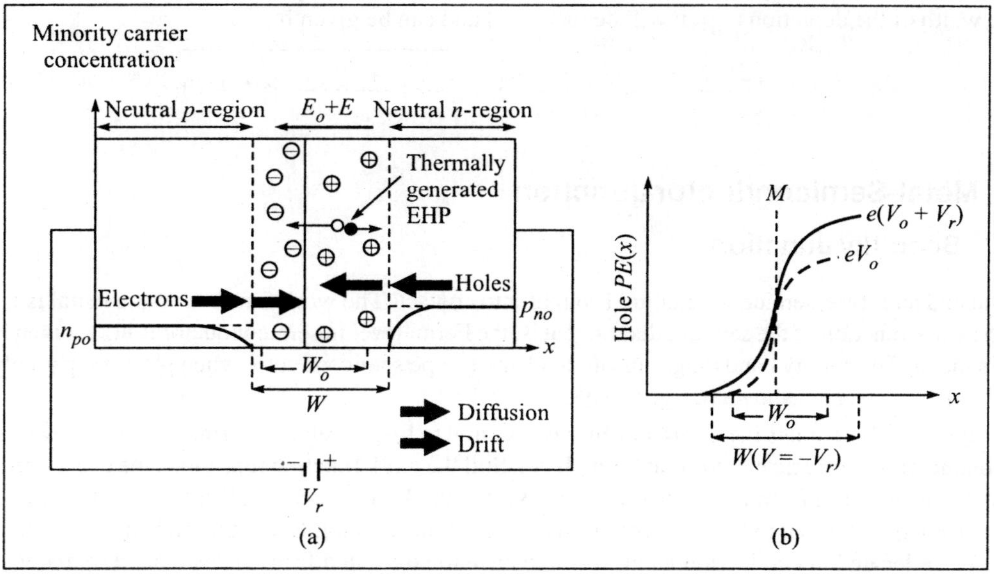

Fig. 6.36 Reversed biased *pn* junction. Minority carrier profile (a); Hole PE across the junction (b).

Considering the effect of thermal generation the equation for reverse current will be modified to

$$J_{rev} = \left(\frac{D_e}{L_e N_a} + \frac{D_h}{L_h N_d}\right) e n_i^2 + \frac{e n_i W}{\tau_g} \tag{6.107}$$

where τ_g is the mean thermal generation time to generate electron-hole pair. And, thermal generation current is

$$J_{gen} = \frac{e n_i W}{\tau_g}$$

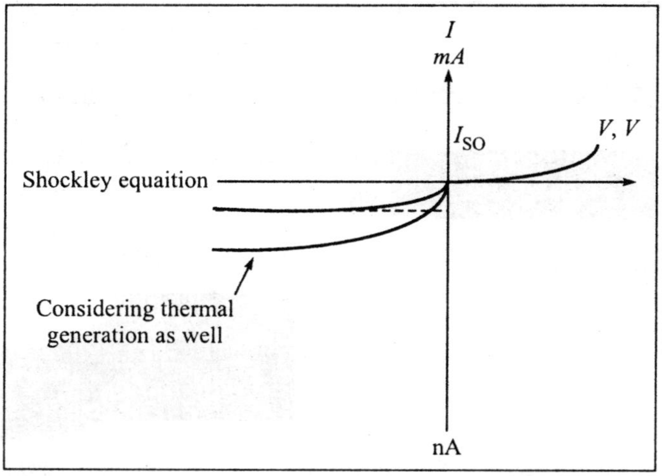

Fig. 6.37 Reverse I-V characteristics of a *pn* junction.

The width of the depletion region will be increased and can be given by

$$W = \left\{ \frac{2\varepsilon(V_0 + V)}{e} \frac{(N_a + N_d)}{N_a N_d} \right\} = \sqrt{\frac{2\varepsilon}{e}\left(\frac{1}{N_a} + \frac{1}{N_d}\right)(V_0 + V)} \quad (6.108)$$

6.9 Metal-Semiconductor Junction

6.9.1 Schottky Junction

A metal and an n-type semiconductor are brought into contact. The work function of the metal is greater than the work function of the semiconductor, that is, the Fermi level in semiconductor is higher than Fermi level in metal. The energy band diagrams of metal and n-type semiconductor when placed separately and brought in to contact are shown in figure 6.38.

In metal work function is the minimum energy required to free the electron from its surface, whereas in semiconductor there are no electrons at Fermi level. Still the work function represents the energy required to remove the electron from the semiconductor. We can think that electron affinity should represent the minimum energy required to free electron from n-type semiconductor. But at thermal equilibrium, not all electrons are located in conduction band, so some have to be excited from valence band to maintain the equilibrium. That is why work function gives the minimum energy required for freeing electron from the semiconductor.

The more energetic electrons in the conduction band of semiconductor can easily tunnel into the metal in search of lower energy levels just above the Fermi level in metal. These electrons from semiconductor pile up on the surface of the metal leaving behind exposed positively charged region with immobile ions, which is depleted of free charge carriers. The depletion region extends into the n-region only. An internal electric field from depleted n-region toward the metal surface near junction where electrons have accumulated

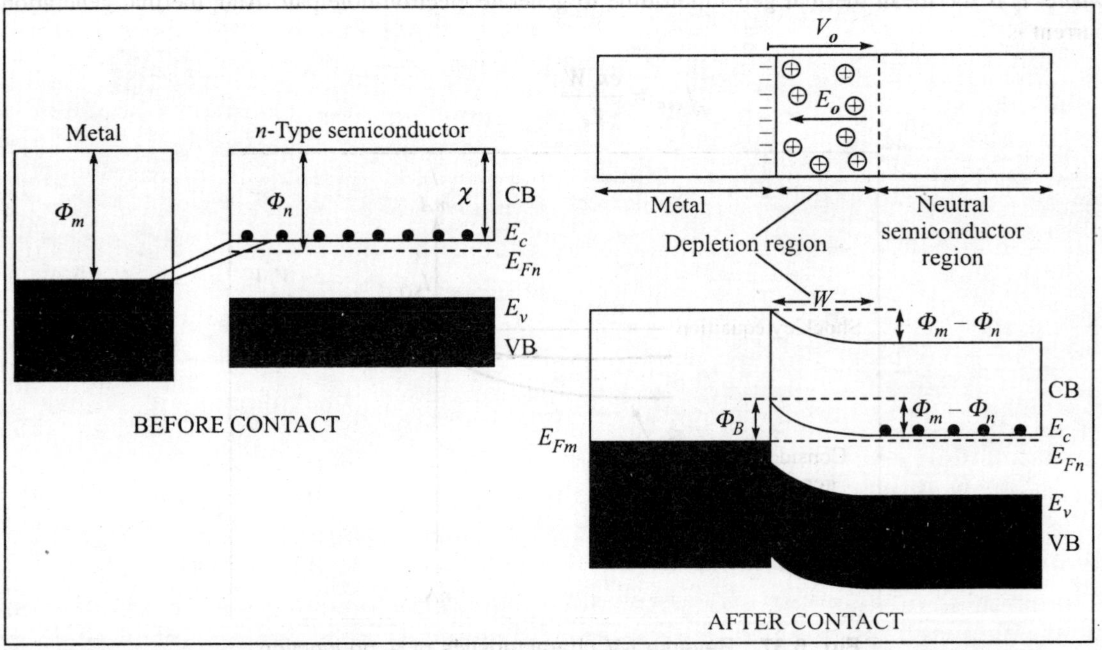

Fig. 6.38 Formation of Schottky junction between metal and n-type semiconductor.

is set up. There is charge difference between the depleted n-region and the metal surface with access electrons, consequently a potential difference known as built-in potential develops between the metal and the semiconductor. The internal electric field will cause drift of electrons from metal surface back to semiconductor. The increase in built-in potential and internal electric field continues till an equilibrium is reached when equal number of electrons cross in both directions. The maximum value of internal field is at the metal-semiconductor junction.

The Fermi level throughout the whole solid (both the metal and semiconductor) must be same. In depletion layer which is depleted of electrons, $E_c - E_{Fn}$ must increase for n to decrease. So, the band bends near the junction to increase $E_c - E_{Fn}$. The bending is just enough for vacuum level to be continuous and changing by $\Phi_m - \Phi_n$ from semiconductor to the metal, as this much energy is required to take an electron from the semiconductor to the metal. The potential energy barrier for electrons moving from metal to the semiconductor is given by $\Phi_B = \Phi_m - \chi = eV_0 + (E_c - E_{Fn})$ is greater than eV_0. Such junction is called Schottky junction and the barrier potential energy as Schottky barrier height.

Under open circuit condition, the number of electrons thermally emitted from metal to semiconductor over potential barrier is equal to the number of electrons thermally emitted over eV_0 from semiconductor to metal. Therefore, no net current flows through the junction. The current density due to electrons thermally emitted from metal to semiconductor is

$$J_1 = A_1 \exp(-\Phi_B/kT) \tag{6.109}$$

where A_1 is a constant with same dimension as the J_1.

Similarly, current due to electrons emitted from conduction band of semiconductor to the metal is given by

$$J_2 = A_2 \exp(-eV_0/kT) \tag{6.110}$$

where A_2 is another constant different from A_1.

During open circuit condition, there will be no current through the junction.

$$J_{oc} = J_2 - J_1 = 0 \tag{6.111}$$

Under forward biased condition, the semiconductor is connected to negative terminal of the battery and the metal is connected to the positive terminal. The whole applied voltage drops in the depletion region owing to its greater resistance compared to the rest. The built-in potential is reduced but the barrier height Φ_B is unchanged. The current due to emission of electron from metal to conduction band of semiconductor over the barrier is unchanged, where as the current due to emission of electrons from conduction band of semiconductor to the metal surface is given by

$$J_2^{for} = A_2 \exp\left[-\frac{e(V_0 - V)}{kT}\right] \tag{6.112}$$

So, the current from semiconductor to metal in forward bias condition considerably increases, whereas current from metal to semiconductor remains the same as in open circuit condition. It is apparent that there will be sustained current from semiconductor to metal given by

$$J = J_2^{for} - J_1 = \exp\left[-\frac{e(V_0 - V)}{kT}\right] - A_2 \exp\left(-\frac{eV_0}{kT}\right)$$

or,

$$J = A_2 \exp\left(-\frac{eV_0}{kT}\right)\left[\exp\left(\frac{eV}{kT}\right) - 1\right]$$

$$J = J_0 \left[\exp\left(\frac{eV}{kT}\right) - 1\right] \tag{6.113}$$

where J_0 is a constant that depends upon material and surface properties of the two solids.

When the Schottky junction is reversed biased, i.e. the positive terminal of battery connected to n-type semiconductor and negative terminal connected to metal, the depletion region will expand as in case of pn junction as the applied field will simply add up to the internal field. The built-in potential will increase from V_0 to $(V_0 + V_r)$. The semiconductor energy band diagram will be shifted down with respect to the metal side because the charge is negative and the voltage is positive. Potential energy barrier for thermal emission of electrons from conduction band of semiconductor to metal increases to $e(V_0 + V_r)$ and the corresponding current will be

$$J_2^{rev} = A_2 \exp\left[-\frac{e(V_0 + V_r)}{kT}\right] \tag{6.114}$$

which will be less than the current due to emission of electrons from metal to semiconductor over potential barrier Φ_B, which is still unchanged from its value at open circuit. So

$$J_2^{rev} = A_2 \exp\left[-\frac{e(V_0 + V_r)}{kT}\right] << J_1 \tag{6.115}$$

Since generally, V_0 is a fraction of volts and the reverse bias is more than a few volts, the reverse bias current is essentially limited by emission current from metal to the conduction band of semiconductor and is very small. The I-V characteristics exhibit rectifying properties and the device is called ***Schottky diode***.

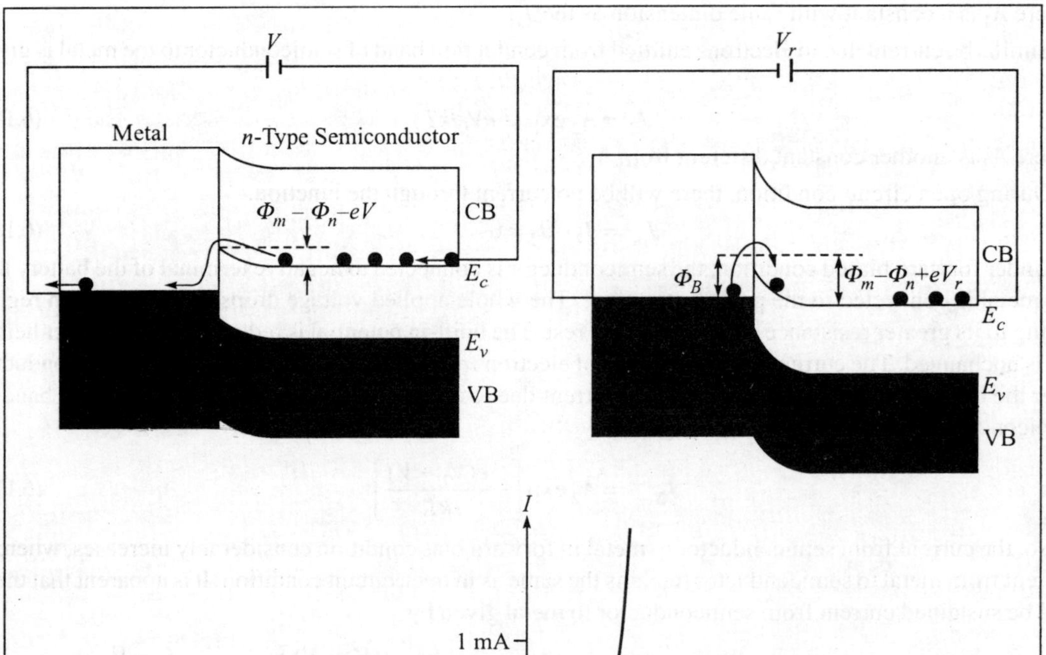

Fig. 6.39 The Schottky junction. Forward biased (a); reverse biased (b); I-V characteristics (c)

6.9.2 Ohmic Contact

An Ohmic contact is a junction between metal and semiconductor that does not limit the current flow. The current is limited by the resistance of the semiconductor outside the contact region. When we say Ohmic contact here it does not mean that the I-V characteristics is linear.

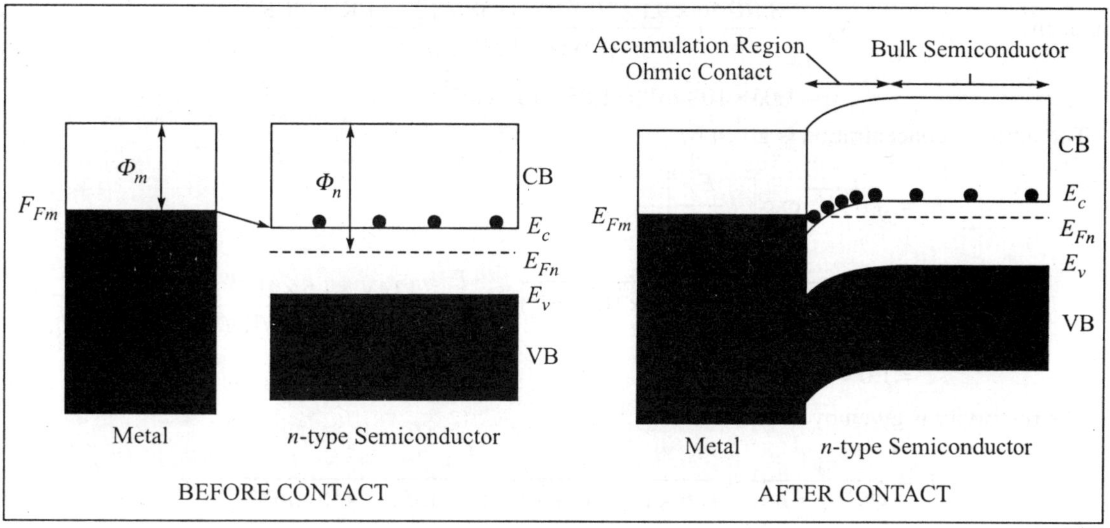

Fig. 6.40 Metal-semiconductor junction with $\Phi_m < \Phi_n$

When a metal and n-type semiconductor are brought together, the electrons in metal near the junction tunnel through to the conduction band of semiconductor because of lower energy level there. Many electrons pile up in conduction band near the junction and equilibrium is reached when further electron cannot tunnel to the conduction band. The energy band is bent downward to decrease $E_c - E_{Fn}$ and increase the electron concentration in conduction band. The electrons in metal and n-region just outside the junction have same energy, so they can cross in any direction under applied field. There is no depletion layer as there are electrons from far left side of metal to far side right of semiconductor. Both the metal and the semiconductor region with electron accumulation have high concentration of electron compared to the bulk of semiconductor. So the current flow is limited by the resistance of the bulk.

6.10 Solved Numerical Examples

1. The density of states related effective masses of electrons and holes in silicon are approximately 1.08 m_e and 0.56 m_e, respectively. The electron and hole drift mobilities at room temperature are 1350 and 450 $cm^2\,V^{-1}\,s^{-1}$ respectively. Calculate intrinsic concentration and intrinsic resistivity of silicon. The energy band gap for silicon is 1.1 eV.

Solution:

Effective densities of states at conduction band edge and valence band edge are given by

$$N_c = \left[\frac{8\pi m_e^* kT}{h^2}\right]^{3/2} \text{ and } N_v = \left[\frac{8\pi m_h^* kT}{h^2}\right]^{3/2}$$

So, $$N_c = \left[\frac{8\pi(1.08\times 9.1\times 10^{-31}\,\text{kg})(1.38\times 10^{-23}\,\text{J/K})(300\,\text{K})}{(6.624\times 10^{-34}\,\text{Js})^2}\right]^{3/2}$$
$$= 2.8\times 10^{25}\,\text{m}^{-3} = 2.8\times 10^{19}\,\text{cm}^{-3}$$

Similarly, $$N_v = \left[\frac{8\pi(0.56\times 9.1\times 10^{-31}\,\text{kg})(1.38\times 10^{-23}\,\text{J/K})(300\,\text{K})}{(6.624\times 10^{-34}\,\text{Js})^2}\right]^{3/2}$$
$$= 1.05\times 10^{25}\,\text{m}^{-3} = 1.05\times 10^{19}\,\text{cm}^{-3}$$

The intrinsic concentration is given by

$$n_i = \sqrt{N_c N_v}\,\exp\left[-\frac{E_g}{2kT}\right]$$

$$n_i = \sqrt{(2.8\times 10^{25}\,\text{m}^{-3})(1.05\times 10^{25}\,\text{m}^{-3})}\,\exp\left[-\frac{1.1\,\text{eV}\times 1.6\times 10^{-19}\,\text{J/eV}}{2(1.38\times 10^{-23}\,\text{J/K})(300\text{K})}\right]$$
$$= 1.0\times 10^{16}\,\text{m}^{-3}$$

The resistivity is given by

$$\rho = \frac{1}{en_i(\mu_e+\mu_h)} = \frac{1}{(1.6\times 10^{-19}\,\text{C})(10^{16}\,\text{m}^{-3})(1350+450)\text{cm}^2\text{V}^{-1}\text{s}^{-1}}$$
$$= 3.5\times 10^3\,\Omega\cdot\text{m} = 3.5\times 10^5\,\Omega\cdot\text{cm}$$

2. Calculate the resistance of pure silicon cubic crystal of 1cm^3 at room temperature. What will be the resistance of the cube when it is doped with 1 arsenic in 10^9 silicon atoms and 1 boron atom per billion silicon atoms? Atomic concentration of silicon is 5×10^{22} cm^{-3}, $n_i = 1.45\times 10^{10}$ cm^{-3}. Assume the mobilities as in the previous example (1).

Solution:

Resistance of the cube is

$$R = \rho\frac{l}{A} = \rho = \frac{l}{\sigma} = \frac{1}{en_i(\mu_e+\mu_h)}$$

$$R = \frac{1}{(1.6\times 10^{-19}\,\text{C})(1.45\times 10^{10}\,\text{cm}^{-3})(1350+450)\text{cm}^2\text{V}^{-1}\text{s}^{-1}}$$
$$= 2.39\times 10^5\,\Omega$$

When the silicon is doped with 1 arsenic atom per 10^9 silicon atoms, then doping concentration is

$$N_d = \frac{N_{Si}}{10^9} = \frac{5\times 10^{22}\,\text{cm}^{-3}}{10^9} = 5\times 10^{13}\,\text{cm}^{-3}$$

At room temperature, all the donors are ionized, so the electron concentration is given by the donor concentration.

$$n \approx N_d = 5\times 10^{13}\,\text{cm}^{-3}$$

The hole concentration then becomes

$$p = \frac{n_i^2}{N_d} = \frac{(1.45 \times 10^{10}\,\text{cm}^{-3})^2}{5 \times 10^{13}\,\text{cm}^{-3}} = 4.2 \times 10^6\,\text{cm}^{-3}$$

Since $p << n$, we can neglect the hole concentration for calculation of resistance, which is given by

$$R = \frac{1}{(1.6 \times 10^{-19}\,\text{C})(5 \times 10^{13}\,\text{cm}^{-3})(1350\,\text{cm}^2\,\text{V}^{-1}\text{s}^{-1})} = 92.6\,\Omega$$

So, there is drastic fall in resistance of the cubic crystal when it is doped with 1 arsenic atom in 10^9 silicon atoms.

Similarly, when the sample is doped with 1 boron atom per billion silicon atoms, the acceptor concentration is again $5 \times 10^{13}\text{cm}^{-3}$ and the resistance of the sample is calculated as, by neglecting the contribution from electrons,

$$R = \frac{1}{(1.6 \times 10^{-19}\,\text{C})(5 \times 10^{13}\,\text{cm}^{-3})(450\,\text{cm}^2\,\text{V}^{-1}\text{s}^{-1})} = 278\,\Omega$$

From this example, we can see the significance of donor doping and acceptor doping, as for the same doping concentrations the resistance of the resulting sample is different.

3. An n-type semiconductor doped with 10^{16} cm^{-3} phosphorous atoms has been doped with 10^{17}cm^{-3} boron atoms. Calculate the electron and hole concentrations in the semiconductor.

Solution:

A semiconductor doped with both the donors and acceptors is called compensated semiconductor. Here pentavalent phosphorous is donor and trivalent boron is acceptor. The resultant acceptor concentration is $N_a - N_d = 10^{17} - 10^{16} = 9 \times 10^{16}$ cm^{-3} The hole concentration is simply given by the resultant acceptor concentration as the intrinsic concentration is very small compared to acceptor concentration.

$$p = N_a - N_d = 10^{17} - 10^{16}\,\text{cm}^{-3} = 9 \times 10^{16}\,\text{cm}^{-3}$$

And, the electron concentration is given by

$$n = \frac{n_i^2}{p} = \frac{(1.45 \times 10^{10}\,\text{cm}^{-3})^2}{9 \times 10^{16}\,\text{cm}^{-3}} = 2.34 \times 10^3\,\text{cm}^{-3}$$

This semiconductor, though having much higher acceptor concentration than donors, is still called compensation-doped semiconductor.

4. An n-type silicon wafer is uniformly doped with 10^{16} antimony atoms per cm^3. Where will be the Fermi level compared to its intrinsic Fermi level? Where will the Fermi level be shifted if the sample is further doped with 2×10^{17} boron atoms per cm^3?

Solution:

An antimony-doped silicon is n-type semiconductor. The intrinsic electron concentration is very small compared to donor concentration. Therefore $n \approx N_d = 10^{16}$ cm^{-3}. The intrinsic concentration and doped concentrations are given by

$$n_i = N_c \exp\left[-\frac{E_c - E_{Fi}}{kT}\right]$$

$$n = N_c \exp\left[-\frac{E_c - E_F}{kT}\right] = N_d$$

So,
$$\frac{N_d}{n_i} = \exp\left[-\frac{E_{Fn} - E_{Fi}}{kT}\right]$$

Solving the above, we have

$$E_{Fn} - E_{Fi} = kT \ln \frac{N_d}{n_i}$$

$$E_{Fn} - E_{Fi} = (1.38 \times 10^{-23}\ \text{JK}^{-1})(300\ \text{K}) \ln \left(\frac{10^{16}\ \text{cm}^{-3}}{1.45 \times 10^{10}\ \text{cm}^{-3}}\right) = 0.348\ \text{eV}$$

When the sample is further doped with boron by acceptor concentration which is greater than already doped donor concentration, the silicon now becomes compensated p-type with resultant hole concentration given by

$$p = N_a - N_d = 2 \times 10^{17} - 10^{16} = 1.9 \times 10^{16}\ \text{cm}^{-3}$$

For intrinsic semiconductor the carrier concentration is, as before, given by

$$n_i = N_v \exp\left[-\frac{E_{Fi} - E_{vi}}{kT}\right]$$

And for doped silicon resulting in p-type semiconductor, the hole concentration is

$$p = N_v \exp\left[-\frac{E_{Fp} - E_v}{kT}\right]$$

So,
$$\frac{p}{n_i} = \exp\left[-\frac{E_{Fp} - E_{Fi}}{kT}\right]$$

After some series of mathematical manipulations we get

$$E_{Fp} - E_{Fi} = -kT \ln \frac{p}{n_i}$$

$$E_{Fi} - E_{Fp} = kT \ln \frac{p}{n_i} = (0.0259\ \text{eV}) \ln \left(\frac{1.9 \times 10^{17}}{1.45 \times 10^{10}}\right) = 0.42\ \text{eV}$$

5. A silicon sample has been doped with 10^{17} arsenic atoms per cm^3. Calculate the conductivity of the sample at 300 K and 400 K. Use the mobility vs impurity concentration graph to determine mobility.

Solution:

To calculate conductivity we can use the donor concentration only since $N_d << n_i$, and $p << n$. The electron drift mobility at $N_d = 10^{17}\text{cm}^{-3}$ is ~800 $\text{cm}^2\text{V}^{-1}\text{s}^{-1}$ at 300 K. So, the required conductivity at 300K is

$$\sigma = en\mu_e = (1.6 \times 10^{-19}\ \text{C})(10^{17}\ \text{cm}^{-3})(800\ \text{cm}^2\ \text{V}^{-1}\ \text{s}^{-1})$$
$$= 12.8\ \text{Sm/cm}.$$

Similarly at 400 K, the drift mobility of electron is 420 $cm^2 V^{-1} s^{-1}$ at the given donor concentration. So, the conductivity of the sample is

$$\sigma = en\mu_e = (1.6\times 10^{-19}\,C)\,(10^{17}\,cm^{-3})\,(420\,cm^2\,V^{-1}\,s^{-1}) = 6.72\ Sm/cm$$

6. What happens to the conductivity if the sample in example 5 is further doped with 9×10^{16} boron atoms cm^{-3}?

Solution:

Still the donor concentration is greater than the acceptor concentration in the sample. So, the sample is n-type. The electron concentration is now reduced to

$$n = N_d - N_a = 10^{17} - 9\times 10^{16} = 10^{16}\ cm^{-3}$$

There is one interesting change associated with the electron drift mobility both at 300 K and 400 K. Due to the crowding of atoms by addition of more impurities (boron atoms), the carriers will face more hindrance to their mobility. So, addition of more impurities although reduces the carrier concentration, but to find mobility we should rather take the sum of the impurity concentration ($1.9\times 10^{17} cm^{-3}$ in this case). The electron mobility at 300 K and 400 K corresponding to this concentration are ~600 $cm^2 V^{-1} s^{-1}$ and 400 $cm^2 V^{-1} s^{-1}$ respectively. So the conductivities at these temperatures are given by

$$\sigma_{300K} = (1.6\times 10^{-19}\,C)\,(10^{17}\,cm^{-3})\,(600\,cm^2\,V^{-1}\,s^{-1}) = 0.96\ Sm/cm$$
$$\sigma_{400K} = (1.6\times 10^{-19}\,C)\,(10^{17}\,cm^{-3})\,(400\,cm^2\,V^{-1}\,s^{-1}) = 0.63\ Sm/cm$$

For both the temperatures, we neglected the conductivity associated with hole as their concentration is very small compared to electron concentration and will not affect the result obtained.

7. Calculate the diffusion coefficient of electrons at 300 K in n-type silicon semiconductor doped with 10^{15} arsenic atoms per cm^3?

Solution:

Using mobility versus dopant concentration graph, we find the electron drift mobility for the given donor concentration is ~1300 $cm^2 V^{-1} s^{-1}$. And using Einstein relation, we have

$$\frac{D_e}{\mu_e} = \frac{kT}{e}$$

$$D_e = \mu_e \frac{kT}{e} = (1300\,cm^2\,V^{-1}\,s^{-1})\,(0.0259\,V) = 33.7\ cm^2\,s^{-1}$$

8. Consider a Schottky junction diode between Al and n-type Si, doped with 10^{16} donors per cm^3. The cross-sectional area of Si is 1m m^2 and electron affinity of silicon is 4.01 eV, work function of Al is 4.28 eV. What will be the theoretical potential barrier height from metal to semiconductor? Using the Richardson's equation for thermionic emission $J_0 = A_0 T^2 \exp(-\Phi_B/kT)$ calculate the reverse saturation current and forward bias current if 0.1V is applied. $A_0 = 3000\ AK^{-2} m^{-2}$. J_0 represents the saturation current density.

Solution:

The theoretical barrier height is

$$\Phi_B = \Phi_m - \chi = 4.28 - 4.01 = 0.27\ eV$$

The typical experimental value is ~0.5 eV due to various reasons such as defects, impurities, dislocations, etc at metal-semiconductor interface. The reverse saturation current is

$$I_0 = A_{J0} A_{A0} T^2 \exp(-\Phi_B/kT)$$
$$I_0 = (0.01\ \text{cm}^2)(30\ \text{AK}^{-2}\ \text{cm}^{-2})(300\ \text{K})^2 \exp(-0.5\ \text{eV}/0.0259\ \text{eV})$$
$$= 1.2 \times 10^{-4}\ \text{A}$$

And the forward current is

$$I_f = I_0\left[\exp\left(\frac{eV_f}{kT}\right) - 1\right] = 1.2 \times 10^{-4}\ \text{A}\left[\exp\left(\frac{0.1\ \text{eV}}{0.0259\ \text{eV}}\right) - 1\right]$$
$$= 5.5 \times 10^{-3}\ \text{A}$$

9. Calculate the built-in potential across the junction formed by joining silicon doped with 10^{17} donor atoms per cm^3 with silicon doped with 10^{16} acceptor atoms per cm^3 at room temperature. Take the intrinsic concentration from previous examples.

Solution:

$$kT/e = 0.0259\ \text{V at 300 K.}$$

$$V_0 = \frac{kT}{e} \ln \frac{N_a N_d}{n_i^2}$$

$$= (0.0259\ \text{V}) \ln\left[\frac{10^{17} \times 10^{16}\ \text{cm}^{-6}}{1.45 \times 10^{10}\ \text{cm}^{-3})^2}\right] = 0.756\ \text{V}$$

10. A heavily doped p-side with acceptor concentration of $10^{18}\ \text{cm}^{-3}$ is connected to n-side with donor concentration of $10^{16}\ \text{cm}^{-3}$. Calculate the built-in potential, depletion width in n-side and p-side, and overall depletion width. The intrinsic concentration is as in previous case and $T = 300$ K.

Solution:

The built-in potential in the junction is calculated as

$$V_0 = \frac{kT}{e} \ln \frac{N_a N_d}{n_i^2}$$

$$= (0.0259\ \text{V}) \ln \frac{(10^{16} \times 10^{18})}{(1.45 \times 10^{10})^2} = 0.816\ \text{V}$$

The depletion width can be calculated by using the following expression

$$W_0 = \sqrt{\frac{2\varepsilon V_0}{e}\left(\frac{1}{N_a} + \frac{1}{N_d}\right)}$$

But here since $N_a >> N_d$, we can simply the formula as

$$W_0 = \sqrt{\frac{2\varepsilon V_0}{e}\frac{1}{N_d}} = \frac{2 \times (11.9 \times 8.85 \times 10^{12}\ \text{Fm}^{-1})(0.816\ \text{V})}{(1.6 \times 10^{-19}\ \text{C})(10^{22}\ \text{m}^{-3})}$$

$$= 3.3 \times 10^{-7}\ \text{m}$$

Now, depletion width in n-region is

$$W_n = \frac{W_0 N_a}{N_a + N_a} = \frac{(3.3\times10^{-7}\,\text{m})(10^{24}\,\text{m}^{-3})}{(10^{22}\,\text{m}^{-3}+10^{24}\,\text{m}^{-3})} = 3.3\times10^{-9}\,\text{m}$$

Similarly, the depletion width in p-region is

$$W_p = W_0 - W_n = 3.3\times10^{-7}\,\text{m} - 3.267\times10^{-7}\,\text{m} = 3.3\times10^{-9}\,\text{m}$$

So almost 99% of depletion layer falls in n-region due to heavily doped p-region where only around 1% depletion layer falls.

11. An abruptp^+n junction diode has a cross-sectional area of 1 mm^2, an acceptor concentration of 5×10^{18} cm^{-3} on p-side and a donor concentration of 10^{16} cm^{-3} on n-side. The lifetime of holes in n-region is 417ns, whereas that of electrons in p-region is 5ns. Mean thermal generation lifetime is about 1ms. The lengths of p and n-sides are 5 micron and 100 micron respectively. Calculate:

(a) Diffusion coefficients and the minority carrier diffusion lengths

(b) Built-in potential across the junction

(c) Current when a forward bias of 0.6 V is applied assuming the current is by the minority carrier diffusion only at 300 K and 373 K and also at 373 K temperature dependence of intrinsic concentration dominates over mobility, diffusion length and diffusion coefficient.

(d) Reverse current when a reverse bias of 5 V is applied.

Solution:

Since we are interested only to the diffusion of minority charge carriers, for an acceptor concentration of $5\times10^{18}\,\text{cm}^{-3}$, the electron mobility (which is minority carrier in p-region) is 120 cm^2 V^{-1} s^{-1} and for a donor concentration of $10^{16}\,\text{cm}^{-3}$, the hole mobility (which is minority charge carrier in n-region) is 440 cm^{2}V^{-1} s^{-1} is obtained from the graph of mobility vs doping concentration. From Einstein relation, we have

$$\frac{D_e}{\mu_e} = \frac{kT}{e} \quad \text{and} \quad \frac{D_h}{\mu_h} = \frac{kT}{e}$$

$$D_e = \mu_e \frac{kT}{e} = (120\,\text{cm}^2\,\text{V}^{-1}\,\text{s}^{-1})\,(0.0259\,\text{eV}) = 3.1\,\text{cm}^2\,\text{s}^{-1}$$

$$D_h = \mu_h \frac{kT}{e} = (440\,\text{cm}^2\,\text{V}^{-1}\,\text{s}^{-1})\,(0.0259\,\text{eV}) = 11.39\,\text{cm}^2\,\text{s}^{-1}$$

Diffusion lengths for electrons and holes are

$$L_e = \sqrt{D_e \tau_e} = \sqrt{(3.1\times\text{cm}^2\text{s}^{-1})(5\times10^{-9}\,\text{s})} = 1.2\times10^{6}\,\text{m}$$

$$L_h = \sqrt{D_h \tau_h} = \sqrt{(11.39\times\text{cm}^2\text{s}^{-1})(417\times10^{-9}\text{s})} = 21.8\times10^{-6}\,\text{m}$$

The built-in potential across the junction is

$$V_0 = \frac{kT}{e} \cdot \ln \frac{N_a N_d}{n_i^2}$$

$$= (0.0259\,\text{V}) \ln \frac{5\times10^{18}\text{cm}^{-3}\times10^{16}\text{cm}^{-3})}{(1.45\times10^{10}\text{cm}^{-3})^{1/2}}$$

$$= 0.856\,\text{V}$$

Assuming the forward current is due to diffusion of minority carriers in neutral regions,

$$I = I_{s0}\left[\exp\left(\frac{eV}{kT}\right) - 1\right] \approx I_{s0}\exp\left(\frac{eV}{kT}\right) \text{ for } V >> \frac{kT}{e}$$

The saturation current I_{s0} is calculated as

$$I_{s0} = AJ_{s0} = A\left[\frac{D_e}{L_e N_a} + \frac{D_h}{L_h N_d}\right] en_i^2 \approx A\,\frac{D_h}{L_h N_d}\,en_i^2 \cdot \text{As } \frac{1}{N_a} << \frac{1}{N_d}$$

$$I_{s0} = \frac{(0.01\,\text{cm}^2)(11.39\,\text{cm}^2\,\text{V}^{-1}\text{s}^{-1})(1.6\times10^{-19}\,\text{C})(1.45\times10^{10}\,\text{cm}^{-3})^2}{(21.8\times10^{-4}\,\text{cm})(10^{16}\,\text{cm}^{-3})}$$

$$= 1.76\times10^{-13}\,\text{A}$$

So, the diode current in forward bias is

$$I = I_{s0}\exp\left(\frac{eV}{kT}\right)$$

$$= 1.76\times10^{-13}\ln\left(\frac{0.6\,\text{V}}{0.0259\,\text{V}}\right)$$

$$2\times10^{-3}\,\text{A} = 2\,\text{mA}$$

At 373 K, the intrinsic concentration is, from graph, equal to $1.2\times10^{12}\,\text{cm}^{-3}$. Reverse saturation current at 373 K can be calculated as

$$I_{s0}(373\,\text{K}) \approx I_{s0}(300\,\text{K})\left[\frac{n_i(373\,\text{K})}{n_i(300\,\text{K})}\right]^2$$

$$= (1.76\times10^{-13}\,\text{A})\left[\frac{1.2\times10^{12}\,\text{cm}^3}{1.45\times10^{10}}\right]^2$$

$$= 1.21\times10^{-9}\,\text{A}$$

And the forward current at this temperature is

$$I(373\,\text{K}) = I_{s0}(373\,\text{K})\exp\left(\frac{eV}{kT}\right)$$

$$I(373\,\text{K}) = (1.2\times10^{-9}\,\text{A})\exp\left[\frac{(1.6\times10^{-19}\,\text{C})\times(0.6\,\text{V})}{(1.38\times10^{-23}\,\text{J/K})(373\,\text{K})}\right]$$

$$= 0.15\,\text{A}$$

When a reverse bias is applied, the depletion width is given by

$$W = \sqrt{\frac{2\varepsilon(V_0 + V_r)}{N_d e}} = \frac{2\times(11.9\times8.85\times10^{-12}\,\text{Fm}^{-1})(0.856 + 5\,\text{V})}{(1.6\times10^{-19}\,\text{C})(10^{22}\,\text{m}^{-3})}$$

$$= 0.88\times1^{-6}\,\text{m} = 0.88\text{ micron}$$

The thermal generation current with reverse bias of 5 V is

$$I_{gen} = \frac{en_i AW}{\tau_g}$$

$$= \frac{(1.6\times10^{-19}\text{C})(1.45\times10^{-10}\text{cm}^{-3})(0.01\text{cm}^2)(0.88\times10^{-4}\text{cm})}{10^{-6}\text{s}}$$

$$= 2\times10^{-9}\text{ A}$$

This current is greater than the reverse saturation current of 1.76×10^{-13} A at 300 K, so the reverse current is dominated by thermal generation current and is 2nA at this condition.

12. A semiconductor is characterized by the energy band diagram given below:

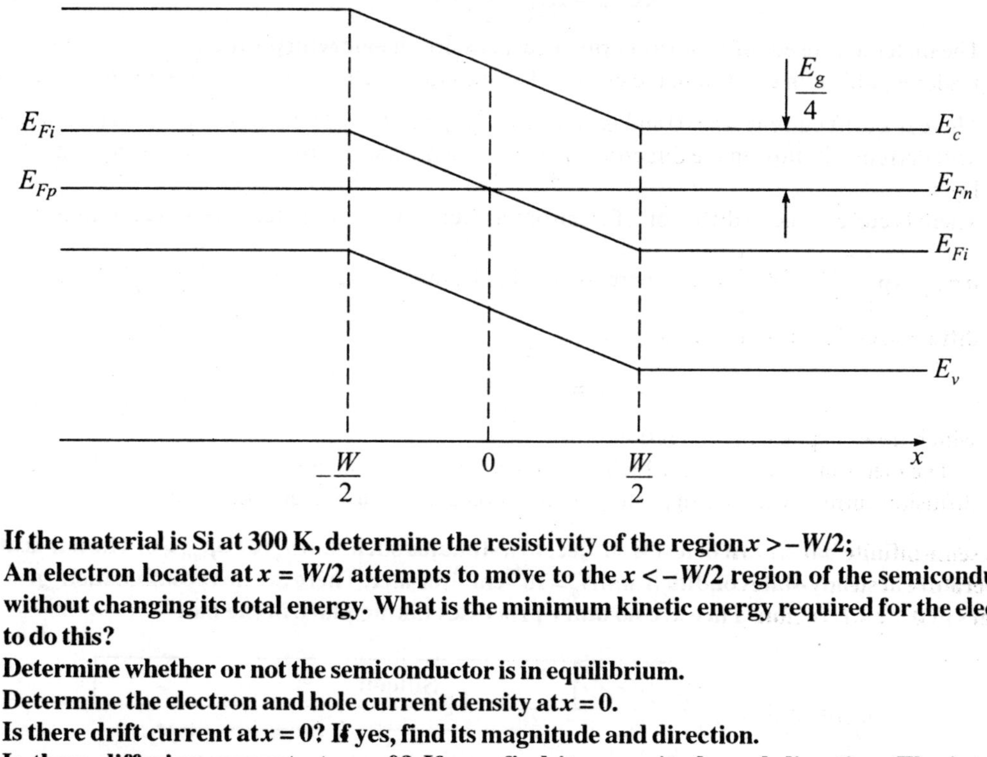

(a) If the material is Si at 300 K, determine the resistivity of the region $x > -W/2$;

(b) An electron located at $x = W/2$ attempts to move to the $x < -W/2$ region of the semiconductor without changing its total energy. What is the minimum kinetic energy required for the electron to do this?

(c) Determine whether or not the semiconductor is in equilibrium.

(d) Determine the electron and hole current density at $x = 0$.

(e) Is there drift current at $x = 0$? If yes, find its magnitude and direction.

(f) Is there diffusion current at $x = 0$? If yes, find its magnitude and direction. The intrinsic concentration at 300 K is $1.18810^{10}\text{cm}^{-3}$ and the $E_g = 1.12$ eV.

Solution:

(a) In the region $x > W/2$, the semiconductor is n-type as the Fermi level is between intrinsic Fermi level and conduction band edge. The electron concentration, and hence donor concentration, in n-region is determined in terms of intrinsic concentration as

$$N_d \approx n = n_i \exp\left[-\frac{E_{Fn} - E_{Fi}}{kT}\right]$$

$$N_d = (1.18\times 10^{10}\,\text{cm}^{-3})\exp\left[-\frac{1.12/4}{0.0259}\right] = 5.58\times 10^{14}\,\text{cm}^{-3}$$

For dopant concentration of 5.85 × 10^{14}cm^{-3}, from the graph mobility versus dopant concentration, electron mobility is found to be ~ 1350 cm^2 V^{-1} s^{-1} which will determine the resistivity of the region.

$$\rho_n = \frac{1}{en\mu_e} = \frac{1}{(1.6\times 10^{-19}\,\text{C})(5.85\times 10^{14}\,\text{cm}^{-3})(1350\,\text{cm}^2\,\text{V}^{-1}\text{s}^{-1})}$$

$$= 7.91\,\Omega\cdot\text{cm}$$

(b) To be able to jump the p region from n region, an electron must possess a kinetic energy at least equal to potential energy barrier between n and p-regions.

$$KE_{\min} = E_c\left(-\frac{w}{2}\right) - E_c\left(\frac{w}{2}\right) = \frac{E_g}{2}$$

(c) The material is in equilibrium as Fermi level does not change with position.

(d) Under equilibrium condition, the current densities due to electrons and holes are zero everywhere.

(e) There is a drift current at $x = 0$ and is given by $J_{e/drift} = en\mu_e\vec{E}$, where n and E both are non zero. The drift current will flow in the direction of electric field which is from n-side to p-side in the depletion layer.

(f) There is current due to diffusion of electrons as well at $x = 0$. The electron concentration is given by $n = n_i \exp\left[\frac{E_{Fn} - E_{Fi}}{kT}\right]$ which increase ingoing from $x = -w/2$ to $x = w/2$. Thus $\frac{dn}{dx} \neq 0$, and electron diffusion current is non-zero at $x = 0$.

$$J_{e/diff} = eD_e\frac{dn}{dx} \neq 0$$

which flows in positive x-direction.

The electron drift current, which is in negative x-direction, must exactly be equal to the electron diffusion current, which is in positive x-direction under equilibrium condition.

13. A semi-infinite bar shown below is subject to a stimulus such that $p_n'(0) = p_{n0}' > 0$. The bar is at room temperature at steady-state condition and $N_d = 10^{14}$cm^{-3} for all x. There are no recombination–generation centers in $0 < x < L$ region. Thee are no other processes taking place in the bar.

Determine the equilibrium electron concentration. Will low level injection exist at $p_{n0}' = 10^{14}$cm^{-3}?

Solution:

All the donor atoms are ionized at room temperature. So the equilibrium electron concentration is given by the donor concentration as

$$n_0 = N_d$$

or, $$n_0 = 10^{14}\,\text{cm}^{-3}$$

Since $p_{n0}' = n_0$, low level injection will not exist.

14. Evaluate the change in conductivity of Silicon specimen when its temperature changes from 300 K to 400 K. Assume usual relationship for temperature dependence of mobility. The effective mass of electron and holes are $1.08m_e$ and $0.56m_e$ respectively. The electron drift mobility and hole drift mobility at 300 K is 1350 cm² V⁻¹ s⁻¹ and 450 cm² V⁻¹ s⁻¹ respectively for the given Si sample. $E_g = 1.1$ eV for Si.

Solution:

We know that the conductivity of Si sample is given by

$$\sigma_{300} = en_{i(300)}\,(\mu_{e(300)} + \mu_{h(300)}) \text{ and}$$

$$\sigma_{400} = en_{i(400)}\,(\mu_{e(400)} + \mu_{h(400)})$$

where $n_{i(300)}$ and $n_{i(400)}$ are the intrinsic concentrations of Si at 300 K and 400 K respectively. Intrinsic concentration in general is given by

$$n_i = \sqrt{N_c N_v}\,\exp\left[-\frac{E_g}{2kT}\right]$$

where

$$N_c = \left[\frac{8\pi m_e^* kT}{h^2}\right]^{3/2} \quad \text{and} \quad N_v = \left[\frac{8\pi m_h^* kT}{h^2}\right]^{3/2}$$

We know that the mobilities are related to the temperature in intrinsic semiconductor by

$$\mu_e \propto T^{-2.3} \quad \text{and} \quad \mu_h \propto T^{-2.2}$$

Now, we can find the ratio of conductivities at 300 K and 400 K as

$$\frac{\sigma_{i(300)}}{\sigma_{i(400)}} = \frac{\left(\frac{8\pi m_e^* kT_{300}}{h^2}\right)^{3/2}\left(\frac{8\pi m_h^* kT_{300}}{h^2}\right)^{3/2}\exp\left(-\frac{E_g}{2kT_{300}}\right)\left[\mu_{e(300)} + \mu_{h(300)}\right]}{\left(\frac{8\pi m_e^* kT_{400}}{h^2}\right)^{3/2}\left(\frac{8\pi m_h^* kT_{400}}{h^2}\right)^{3/2}\exp\left(-\frac{E_g}{2kT_{400}}\right)\left[\mu_{e(400)} + \mu_{h(400)}\right]}$$

$$\frac{\sigma_{i(300)}}{\sigma_{i(400)}} = \frac{T_{300}{}^3 \exp\frac{E_g}{2K}\left(\frac{1}{T_{400}} - \frac{1}{T_{300}}\right)\left[\mu_{e(300)} + \mu_{h(300)}\right]}{T_{400}{}^3\left[\mu_{e(400)} + \mu_{h(400)}\right]}$$

$$\frac{\sigma_{e(400)}}{\sigma_{e(300)}} = \left[\frac{T_{400}}{T_{300}}\right]^{-2.3}$$

So, $$\mu_{e(400)} = \mu_{e(300)}\left[\frac{T_{400}}{T_{300}}\right]^{-2.3} = 1350\times\left[\frac{400}{300}\right]^{-2.3} = 696.6 \text{ cm}^2\,\text{V}^{-1}\,\text{s}^{-1}$$

Similarly, the hole mobility at 400 K can be calculated as

$$\mu_{h(400)} = \mu_{e(300)}\left[\frac{T_{400}}{T_{300}}\right]^{-2.2} = 450\times\left[\frac{400}{300}\right]^{-2.2} \approx 239 \text{ cm}^2\,\text{V}^{-1}\,\text{s}^{-1}$$

Now, $$\frac{\sigma_{i(300)}}{\sigma_{i(400)}} = \frac{300^3 \exp\left[\left(\frac{1.1\times 1.6\times 10^{-19}}{2\times 1.38\times 10^{-23}}\right)\left(\frac{1}{400}-\frac{1}{300}\right)\right][1350+450]\times 10^{-4}}{400^3\,[696.6+239]\times 10^{-4}}$$

$$\sigma_{i(300)} = 3.995\times 10^{-3}\times \sigma_{i(400)}$$

Then we can write,

$$\sigma_{i(400)} = 250.31\times \sigma_{i(300)}$$

The intrinsic conductivity of the given Si sample increases by around 250 times compared to the conductivity at 300 K when temperature reaches 400 K.

15. How the conductivity of the donor doped Si changes as temperature is reduced from 250 K to 200 K. The doping concentration is 10^{25}m^{-3}.

Solution:

Conducyivity of donor-doped Si is given by

$$\mu_n \approx en\mu_e$$

where $n \approx N_d = 10^{25}$ m^{-3}

Now, the electron drift mobility for the doping concentration of 10^{25}m^{-3} in the given temperature changes positively with the change in temperature, which is due to impurity scattering domination.

From the experimental plot of drift mobility vs. temperature for the given doping concentration, we have

$$\mu_{200} = 93 \text{ cm}^2\text{ V}^{-1}\text{ s}^{-1}$$
$$\mu_{250} = 100 \text{ cm}^2\text{ V}^{-1}\text{ s}^{-1}$$

Now, the conductivities can be evaluated as

$$\frac{\sigma_{250}}{\mu_{200}} = \frac{100}{93} = 1.075$$

Significant to note is that, a change of temperature by almost a quarter causes change in conductivity by merely 7.5%. This is mainly due to saturation of the ionization in the doped semiconductor.

From the same graph of electron drift mobility vs. temperature for n-type semiconductor, one might conclude that as mobility is seen decreasing in most of the cases with increase in temperature, the conductivity might be actually decreasing. This is true but for only a narrow temperature range only. For all other cases, the conductivity of doped semiconductor will be increasing with the rise in temperature as the semiconductor will start behaving like intrinsic above a certain temperature and any reduction in mobility will be compensated by increase in carrier concentration both of which affect the conductivity. The same is verified in problem no. (6) above.

Semiconductor Materials Processing

7 Chapter

7.1 Crystal Growing

Silicon and germanium are two of the most widely used semiconductor materials. Of these two, silicon has been dominant since it was first used in early 1950s. The widespread use of Si is based on various favorable factors attached with the material.

- It is most abundant element after oxygen in the earth. Around 28% of earth is composed of silicon.
- The raw material from which pure silicon is extracted is found everywhere in the nature. So, it is very cheap.
- The natural oxide of silicon SiO_2 is an excellent insulating medium.
- It has good heat conductivity, which is also favorable for many applications.
- The energy band gap is enough large to be used in various applications. But not so large to cancel out their usefulness.
- It is non-toxic. So, no special care is needed when using Si for industrial and commercial applications.

Pure silicon is required for various applications. In nature, there is no pure silicon. It can be found in compound form only. So there are certain procedures to obtain pure silicon as explained below.

- SiO_2 is heated to high temperature in a reducing environment to obtain approximately 98% pure silicon powder.
- The silicon powder is then reacted with HCl gas to form trichlorosilane gas.

$$Si + 3HCl \rightarrow SiHCl_3 + H_2$$

  which is then distilled and reduced with hydrogen to form polycrystalline silicon (reverse of above process).
- The polycrystalline silicon is then melted and cast into ingots, from which the single crystal is obtained.

There are basically two methods of obtaining pure silicon crystals:

(i) Floating zone refining and crystal growth
(ii) Czochralski growth.

7.1.1 Floating Zone Method of Crystal Growth

This method, in addition to, being capable of producing single crystals and also removing the extra impurities contained in the semiconductor material. The impurities can be removed very easily as they prefer to go to the liquid phase.

This type of crystal growing material is rarely used because of the high temperature requirement of silicon, there is always possibility of silicon crystals adhering to their crucible wall which causes deviation from the perfect lattice structure required for the silicon crystal.

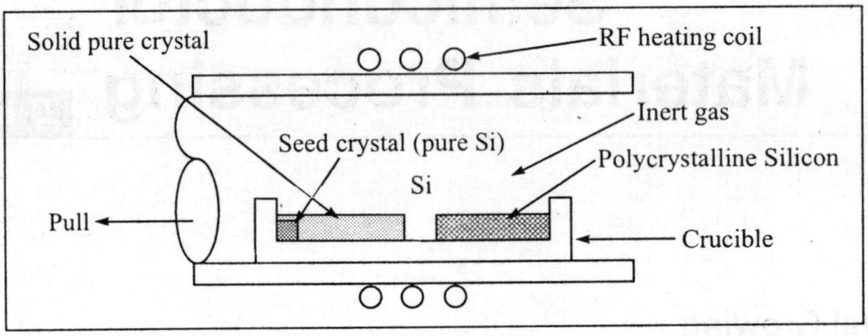

Fig. 7.1 Floating zone method of crystal growing

The RF (Radio-frequency) heating coils heats the crucible in which the polycrystalline silicon is placed. As the Si is heated, it melts at high temperature and by solidification by slowly pulling it to the left where there is seed crystal; single pure crystal for IC fabrication is obtained. The inert gas is used to suppress evaporation and prevent oxidation.

7.1.2 Czochralski Growth

Large, single, Si crystals are grown by Czochralski method, which involves growing a single crystal ingot from the melt using solidification on a seed crystal. Molten Si is held in quartz crucible in a graphite susceptor, which is heated by a RF heating coil. The seed crystal is lowered to touch the melt and then

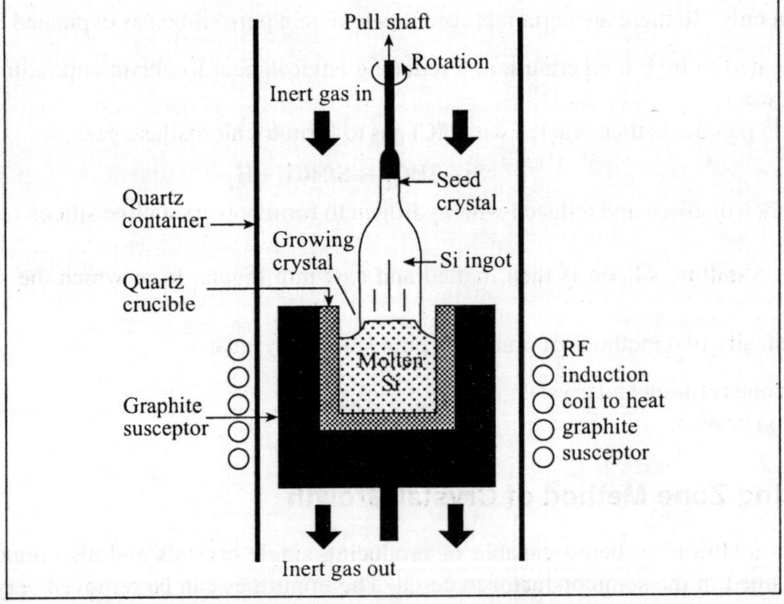

Fig. 7.2 Single crystal growth by Czochralski method

slowly pulled out of the melt. In solidification pure crystal grows. The seed is rotated during pulling stage, to obtain a cylindrical ingot. To suppress evaporation from the melt and prevent oxidation inert gas is used.

Initially, as the crystal is withdrawn, its cross-sectional area increases; it then reaches a constant value determined by the temperature gradients, heat losses, and rate of pull. As the melt solidifies on crystal, heat of fusion is released and must be conducted away, otherwise it will raise the temperature of the crystal and re-melt it.

The above mentioned processes will result is single crystal rods being formed. These rods are sliced by a diamond band saw, etched and polished to obtain wafers that are 0.3 to 0.4 mm thick. These wafers are the basic building blocks for device fabrication.

7.2 Diffusion System

7.2.1 Diffusion

This is one of the primary method of introducing impurities such as boron, phosphorous, antimony into silicon to control the majority-carrier type and sheet resistivity of layers formed in the wafer. Diffusion process begins with the deposition of a high concentration of the desired impurity on the silicon surface through windows etched in the protective barrier layer. At high temperatures (900 – 1200°C), the impurity atoms move from the surface into the silicon crystal by substitutional or interstitial diffusion mechanisms. In substitutional diffusion, the impurity atom hops from one crystal lattice site to another. The impurity atom substitutes for a silicon atom in the lattice. There must be vacancies in the lattice for this process to go on, and the vacancies generally exist in the lattice. At high temperatures, the silicon atoms are displaced from their usual positions to interstitial space between lattices, and vacancies are created. The substitutional diffusion process in which impurities are introduced and silicon atoms are displaced into interstitial sites is called interstitialcy diffusion.

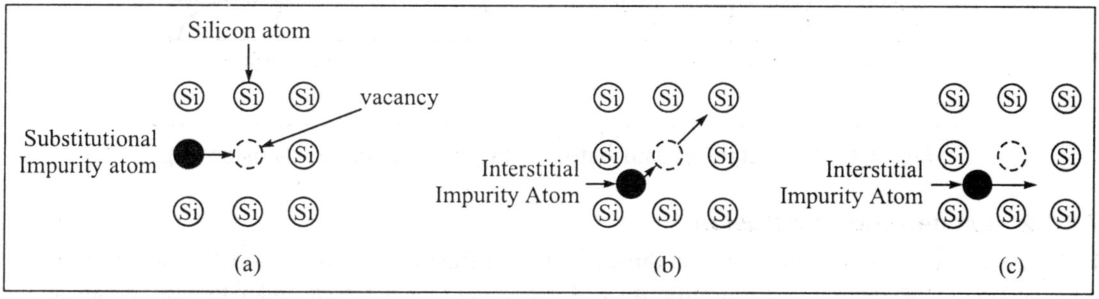

Fig. 7.3 Two dimensional Si lattice. Substitutional diffusion (a); Interstitialcy, in which impurity atom replaces the Si atom (b); Interstitial diffusion in which impurity atom does not replace Si atom in crystal lattice (c).

The mathematical expression of diffusion process in 1-dimension can be derived using Fick's laws of diffusion. Fick's first law of diffusion states that the particle flow per unit area (particle flux), J, is directly proportional to concentration gradient of the particle.

$$J = -D\frac{\partial N}{\partial x} \tag{7.1}$$

where N is the particle concentration and D is the diffusion coefficient. The –ve sign indicates that particles move from region of high concentration to low concentration.

The continuity equation for particle flux is

$$\frac{\partial N}{\partial t} = -\frac{\partial J}{\partial x} \tag{7.2}$$

Combining equations (7.1) and (7.2) gives Fick's 2nd law of diffusion:

$$\frac{\partial N}{\partial t} = D\frac{\partial^2 N}{\partial x^2} \tag{7.3}$$

Two specific boundary conditions, namely, constant source diffusion and limited-source diffusion are important in modeling the diffusion.

7.2.1.1 Constant-source diffusion

During constant source diffusion, the impurity concentration is held constant at the surface of wafer. As time progresses, the diffusion front proceeds further and further into the wafer with the surface concentration remaining constant. The number of impurity atoms per unit area in Si called dose, Q, increases with time, and an external impurity source must supply a continual flow of impurity atoms to the surface of wafer.

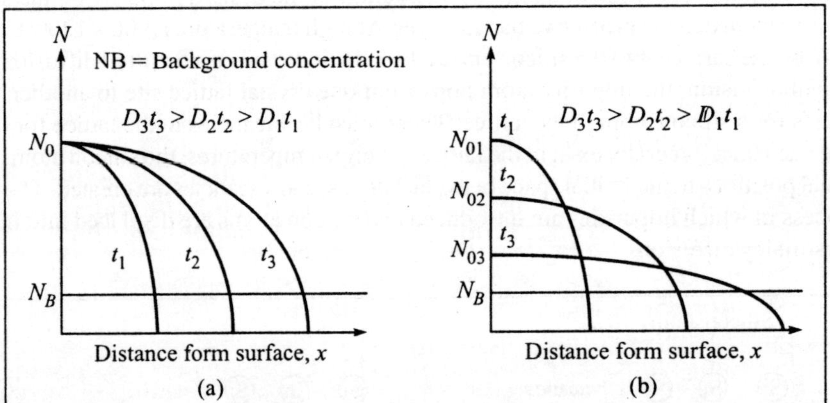

Fig. 7.4 A constant source diffusion (a); limited source diffusion (b).

7.2.1.2 Limited source diffusion

In this process the dose remains constant throughout the diffusion process. As the diffusion front moves into the wafer, the surface concentration must decrease so that the area under the curve can remain constant with time.

A short constant source diffusion is often followed by a limited source diffusion, resulting in a two step diffusion process. The constant course diffusion is used to establish a known dose in a shallow layer on the surface of the silicon and is called the predeposition step. The second diffusion called the drive-in step is used to move the diffusion front to the desired depth.

7.2.1.3 Junction formation by vertical diffusion

The goal of most diffusions is to form pn junctions by converting p-type material to n-type and vice versa. The point at which the diffused impurity profile intersects the back-ground concentration N_B is called metallurgical junction depth x_j. The net impurity concentration at x_j is zero.

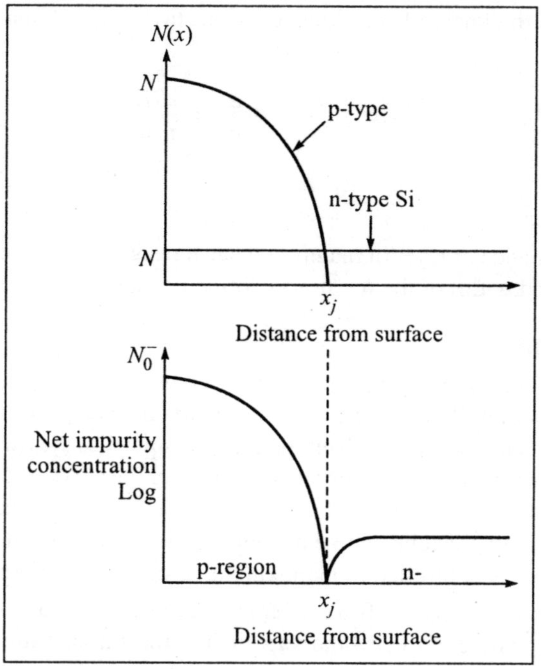

Fig. 7.5 Formation of pn junction. P type diffusion into n type wafer (a); Net impurity concentration in the wafer (b).

7.2.1.4 Lateral diffusion

During diffusion, impurities not only diffuse vertically but also move laterally under the edge of any diffusion barrier. Lateral diffusion is an important effect coupling device and process design and was an important factor driving the development of self-aligned polysilicon-gate MOS processes.

7.2.1.5 Sheet resistance

In diffused layers, resistivity is a strong function of depth. For circuit and device design, it is more convenient to express sheet resistance, which is given by equation (7.4).

$$R = \rho\frac{L}{A} \tag{7.4}$$

where ρ = resistivity of the material,
L = length
A = cross-sectional area of the material

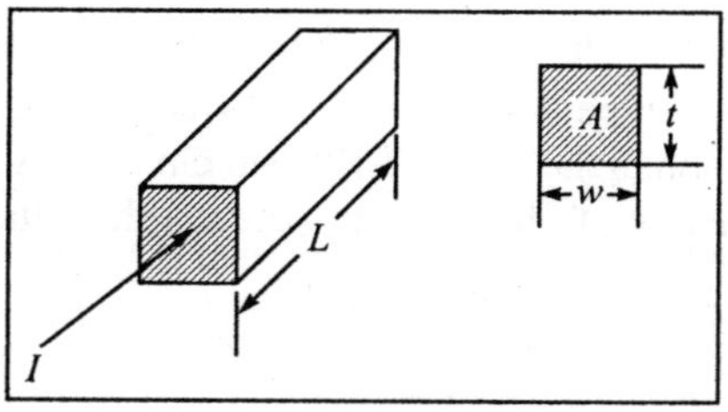

Fig. 7.6 Resistance block of material having uniform resistivity.

Using w as width and t as thickness of the block, cross-sectional area $A = wt$ and the resistance of the block can be written as

$$R = \rho\frac{L}{A} = \rho\frac{L}{wt} = \frac{\rho}{t}\frac{L}{w}$$

$$R = R_s\frac{L}{w} \qquad (7.5)$$

where R_s is called sheet resistance of layer of material. If sheet resistance is given, a circuit designer needs to specify only the length and width of the resistor to define its value.

7.2.1.6 Diffusion systems

The open-furnace-tube system using solid, liquid, or gaseous sources is the most common diffusion technology used in IC production. Wafers are placed in a quartz boat and positioned in the center zone of the furnace, where the wafers are heated to high temperatures. Impurities are transported to silicon surface, where they diffuse into the wafer. In solid source system, carrier gases N_2 or O_2 flow at a controlled rate over a source boat placed in the furnace tube as shown in fig.7.7.a. The carrier gas picks up the vapour from the source and transports it down the tube, where the dopant species is deposited on the surface of wafer. The temperature of the source is controlled to maintain the desired vapour pressure.

In liquid source system shown in fig.7.7.b, a carrier gas passes through a bubbler where it picks up the vapour of the liquid source. The gas carries the vapour into the furnace tube where it reacts with the surface of the silicon wafer.

In gas source system, the dopants are directly supplied to the furnace tube in the gaseous state. The common gas source are extremely toxic, and additional input purging and trapping systems are required to ensure that all the source gas is removed from the system before wafer entry or removal.

B, P, As and Sb are the most common dopants used diffusion. Of these, B is the most popular p-type dopant due to its high solubility in silicon and high surface concentration of 4×10^{20} per cm^3. Also, the lemental boron is inert up to temperatures exceeding the melting point of silicon. A surface reaction with boron trioxide is used to introduce boron into the silicon surface: $2B_2O_3 + 3\text{Si} \leftrightarrow 4B + 3\text{Si}O_2$ (7.6)

Common solid source of boron are trimethylborate(TMB) and boron nitrate wafers. TMB vapour reacts in furnace tube with oxyzen to form boron trioxide:

$$2\,(CH_3O)_3\,B + 9O_2 \xrightarrow{900°C} B_2O_3 + 6CO_2 + 9\,(H_2O) \qquad (7.7)$$

The most common liquid source of boron is boron tribromide. The reaction is:

$$4\,BBr_3 + 3O_2 \rightarrow 2\,B_2O_3 + 6\,Br_2 \qquad (7.8)$$

The primary gaseous source of boron is diborane. The reaction is:

$$B_2H_6 + 3O_2 \xrightarrow{300°C} 4B_2O_3 + 3H_2O \qquad (7.9)$$

or

$$B_2H_6 + 6CO_2 \rightarrow B_2O_3 + 6CO_2 + 3H_2O \qquad (7.10)$$

Phosphorous which is common n-type dopant used in diffusion has higher solubility in silicon than boron and has a surface concentration in the range of 10^{21} per cm^3 during high temperature diffusion. P is introduced into Si through the reaction of phosphorous pentaoxide at the wafer surface.

$$2P_2O_5 + 5Si \leftrightarrow 4P + 5SiO_2 \qquad (7.11)$$

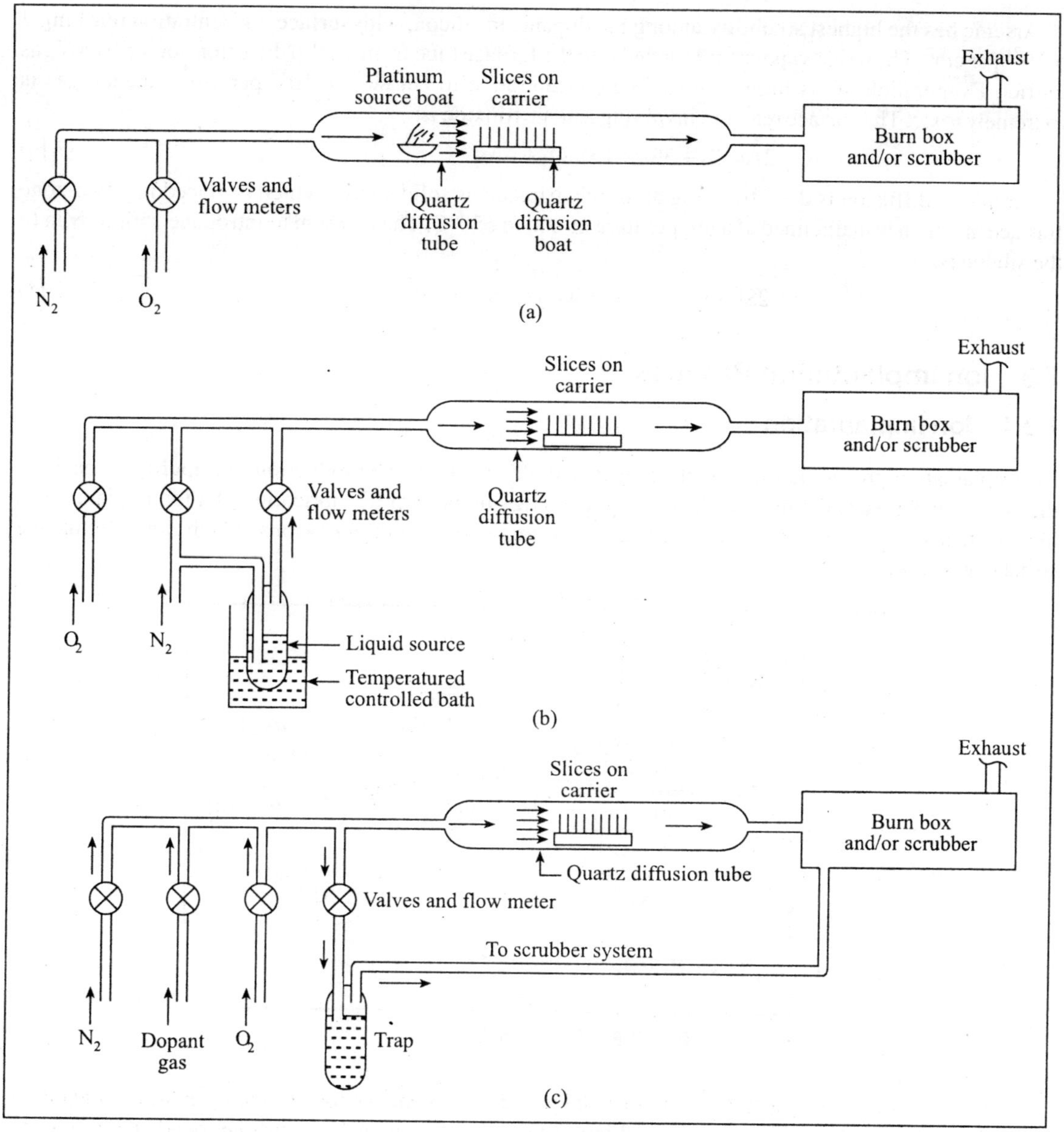

Fig. 7.7 Open furnace tube diffusion system. Solid impurity source (a), liquid impurity source (b) and gaseous impurity source (c)

Solid wafers of phosphorous pentaoxide can be used as source for phosphorous. The reaction to obtain P_2O_5 by using phosphorous oxychloride $POCl_3$ and oxygen is:

$$4POCl_3 + 3O_2 \rightarrow 2P_2O_5 + 6Cl_2 \tag{7.12}$$

Phosphine is used as gaseous source for phosphorous.

$$2PH_3 + 4O_2 \rightarrow P_2O_5 + 3H_2O \tag{7.13}$$

Arsenic has the highest solubility among the dopants in silicon, with surface concentration reaching 2 $\times 10^{21}$ per cm^5. The oxide vapours are carried into the furnace tube from a solid diffusion source by a *N* gas carrier. Evaporation of As limits the surface concentration to below 3 $\times$ 10^{19} per cm^5. Arsenic gas is extremely toxic. The surface reaction involving arsenic trioxide is:

$$2As_2O_3 + 3Si \leftrightarrow 3SiO_2 + 4As \tag{7.14}$$

Antimony diffusion is done by using antimony trioxide as solid source which is placed in a two-zone furnace in which is maintained at a temperature of 600 to 650°C. The reaction to introduce antimony in to the silicon is:

$$2Sb_2O_3 + 3Si \rightarrow 3Si\,O_2 + 4Sb \tag{7.15}$$

7.3 Ion Implantation Process

7.3.1 Ion Implantation

Ion Implantation process has many advantages over diffusion for the production of impurity atoms in to the silicon wafer, so is the most widely used technology in modern IC fabrication. An ion implanter is a high-voltage particle accelerator producing a high-velocity beam of impurity ions, which can penetrate the surface of Si wafers.

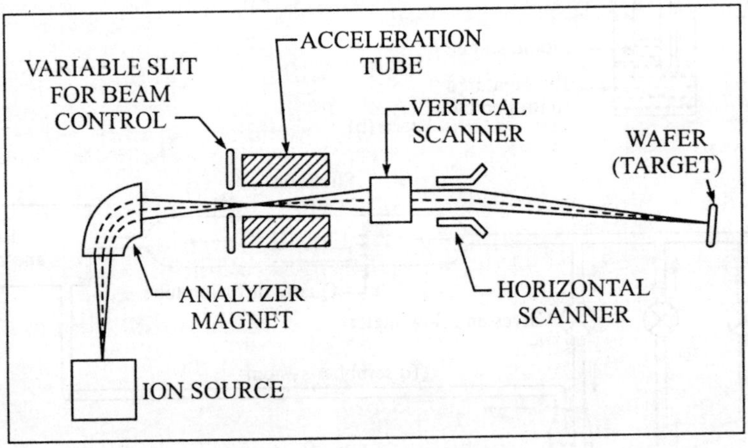

Fig. 7.8 Typical Ion Implanter

The ion source, which operates at high voltage (25kV), is meant for producing plasma containing desired impurity, but it produces undesired impurities as well. An analyzer magnet bends the ion beam through a right angle to select the desired impurity ion, which then passes through an aperture slit into the main accelerator column. The main function of accelerator column is to add energy of up to 175 keV to the beam and accelerate the ions to their final velocity. Scanning system consists of x- and y-axis deflection plates to scan the beam across the wafer to give a uniform implantation and to build up the desired dose. The beam is bent slightly to prevent neutral particles hitting the target. The target chamber has silicon wafers, which will be targeted by the impurity ions.

A charged particle with velocity ***v*** in magnetic filed ***B*** will experience a force ***F*** given by

$$F = q\,(vxB) \tag{7.16}$$

The force will tend to move the particle in a circle, and the centrifugal force will balance F. But, qvB = mv^2/2 and mv^2/2 =qV. So the magnitude of magnetic field B can be adjusted to select an ion spice with a given mass:

$$|B| = \sqrt{(2mV/qr^2)} \tag{7.17}$$

where V – acceleration voltage,

v – velocity of particle,

The ion source operates at constant potential, so voltage is known. So the ion spice is selected by changing *dc* current supplying the magnetic analyzer. The target chamber is maintained at relatively low temperatures during the implantation, which prevents undesired spreading of impurities by diffusion – very important in VLSI fabrication. Also, by ion implantation wide range of impurities can be implanted compared to that of diffusion. The main drawback is the cost of the ion implanter.

Ion implantation usually follows a Gaussian distribution as given by:

$$N(x) = N_P \exp\left[-\frac{(x-R_P)^2}{2\Delta R_P{}^2}\right] \tag{7.18}$$

where $N(x)$ is the impurity concentration,

N_P is the peak concentration,

R_P – the projected range

ΔR_P - standard deviation called straggle

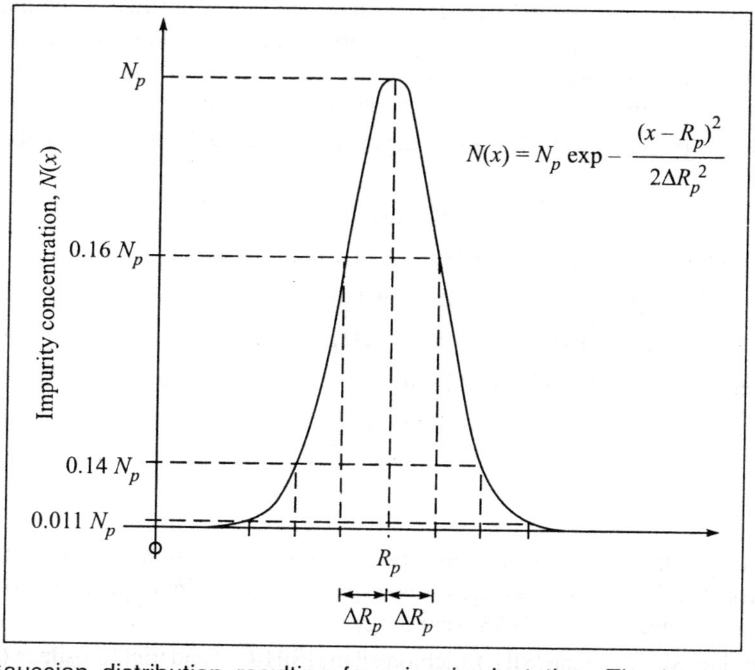

Fig. 7.9 Gaussian distribution resulting from ion implantation. The impurity is shown implanted completely below the wafer surface (x = 0).

The ion implantation is often used to form shallow pn junctions for various device applications where the junction depth can be found by equating the implanted distribution function to the background concentration N_B as:

$$N_P \exp\left[-\frac{(x_j - R_P)^2}{2\Delta R_P{}^2}\right] = N_B$$

or $$x_j = R_P \pm \Delta R_P \sqrt{2\ln(N_P/N_B)} \qquad (7.19)$$

7.3.2 Lattice Damage and Annealing

During the implantation process, an ion impact can knock atoms out of Si lattice, damaging the implanted region in the crystal. For higher doses, the implanted layer becomes amorphous. With heavier impurities lower doses will be required to create an amorphous layer.

Implantation damage can be removed by annealing. After the implantation, the wafer is heated to a temperature ~ 900°C for approximately 30 minutes. At this temperature, Si atoms can move back into lattice sites, and impurity atoms can enter substitutional sites. After annealing cycle, nearly all of the implanted dose becomes electrically active, except for impurity concentration exceeding 10^{19} cm^{-3}. Truly amorphous layers can be annealed at lower temperatures through solid-phase epitaxy.

7.3.3 Chemical Vapour Deposition

Chemical vapour deposition (CVD) forms thin films on the surface of a substrate by thermal decomposition and/or reaction of gaseous compound. The desired material is deposited directly from the gas phase onto the surface of substrate. Polysilicon, silicon dioxide and silicon nitride are deposited using CVD technique. CVD can be performed at pressures for which the mean free path for gas molecules is quite small.

Si is deposited in low pressure chemical vapour deposition (LPCVD) by using thermal decomposition of silane.

$$SiH_4 \rightarrow Si + 2H_2 \qquad (7.20)$$

Low-pressure systems use either 100% or 25-30% silane diluted with nitrogen.

Silicon dioxide can be deposited by using silane and oxygen:

$$SiH_4 + O_2 \rightarrow SiO_2 + 2H_2 \qquad (7.21)$$

Phosphorous pentaoxide can be deposited by using phospine and oxygen as:

$$4PH_3 + O_2 \rightarrow 2P_2O_5 + 6H_2 \qquad (7.22)$$

7.3.4 Epitaxial Growth

Epitaxial growth comes from Greek word "epitaxy" meaning "arranged upon". The growth of a signle crysattline layer from the vapour phase is called vapour-phase epitaxy. Liquid-phase epitaxy and molecular-beam epitaxy are also widely used in GaAS technology.

Silicon tetrachloride ($SiCl_4$), Silane (SiH_4), dichlorosilane (SiH_2Cl_2) and trichlorosilane ($SiHCl_3$) are most commonly used for silicon vapour phase epitaxy.

$$SiCl_4\,(\text{gas}) \leftrightarrow Si\,(\text{solid}) + 4HCl\,(\text{gas}) \qquad (7.23)$$

The reaction takes place at approximately 1200°C and is reversible.

Epitaxial growth can also be achieved by pyrolytic decomposition of silane:

$$SiH_4 \xrightarrow{650°C} Si + 2H_2 \tag{7.24}$$

The reaction is not reversible and takes place at low temperature.

In liquid-phase epitaxy, the substrate is brought into contact with a solution containing the material to be deposited in liquid form. The substrate acts as a seed for material crystallizing directly from the solute.

In liquid phase epitaxy, the substrate is brought into contact with a solution containing the material to be deposited in liquid form. The substrate acts as seed for material crystallizing directly from the solute.

In molecular beam epitaxy, the crystalline layer is formed by deposition from a thermal beam of atoms or molecules in ultrahigh vacuum conditions. Substrate temperature during this process ranges from 400-900°C.

7.4 Monolithic IC Fabrication: Planar Process

In the planar process, the fabrication of the IC is done plane by plane at one surface of the wafer so that the deepest doped region tends to be fabricated first. We will consider fabrication of two neighbouring integrated npn bipolar junction transistors. At first, heavily doped n^+ buried layers of all npn BJTs are fabricated. For this, a silicon wafer, in the form of thin p-type single Si-crystal, is oxidized in an oxidation furnace to form a thin layer of SiO_2, which is excellent barrier against diffusion. The silicon dioxide layer adheres to the Si crystal without any cracks and pores and forms a continuous layer with an amorphous structure. SiO_2 can be easily etched to open windows to allow diffusion of dopants through selective areas. A process called lithography, which literally means writing on stone, provides a means for defining windows in the oxide. The surface of the oxidized wafer is coated with a thin photoresist emulsion. A positive photoresist, which becomes chemically soluble when exposed to UV light will be considered here. A black and white photomask is fabricated in which the white areas correspond to regions where diffusion windows are needed in SiO_2. The wafer is illuminated by UV light, which causes photoresist regions under transparent areas of photomask become soluble by UV light. Then the photoresist is developed in a suitable chemical solution in which soluble regions are chemically washed away and only insoluble region remain. The photoresist mask that remains on the oxide is hardened to resist etchant to be used for oxide removal. Windows re then etched in oxide by using an etchant (HF acid). The process in which UV light is used to reproduce the diffusion window pattern is called photolithography. Once Si crystal surface has been accessed through the diffusion windows, donor dopants are diffused into crystal. The surface region where diffused donors exceed the bulk acceptor concentration becomes n-type by compensation doping. By suitable temperature and pressure, heavy n-type doping is done to form n^+ doping. The SiO_2 used in n^+ buried layers diffusion is removed by etching, and then a thin lightly doped epilayer is grown on the surface. The entire diffused and doped region occur in the epilayer. Individual components must be isolated from each other by embedding each component in n-type island and p-type island. The whole crystal is reoxydized and using photolithography next set of window patterns are defined. Windows are then etched around each collector region (fig 7.10.i). Acceptor impurities are then diffused to the substrate to form p^+ regions around each collector (fig 7.10.j). The whole crystal is again reoxydized and the new oxide grows on the surface of the already present oxide as well as Si crystal (fig7.10k). Photolithography is used once again and diffusion windows are then etched (7.10.l). Boron gas is diffused through the windows to form a lightly p-type base (7.10m). Once again using photolithography new windows are etched in the oxide where n^+ emitter region and n^+ ohmic contacts to the collector are to be formed (7.10n). Donor dopants are then introduced to form n^+ emitter regions and n^+ contact regions of the collector (7.10.o). The bipolar fabricated at the end of step o need contacts and interconnections. For this the whole surface is

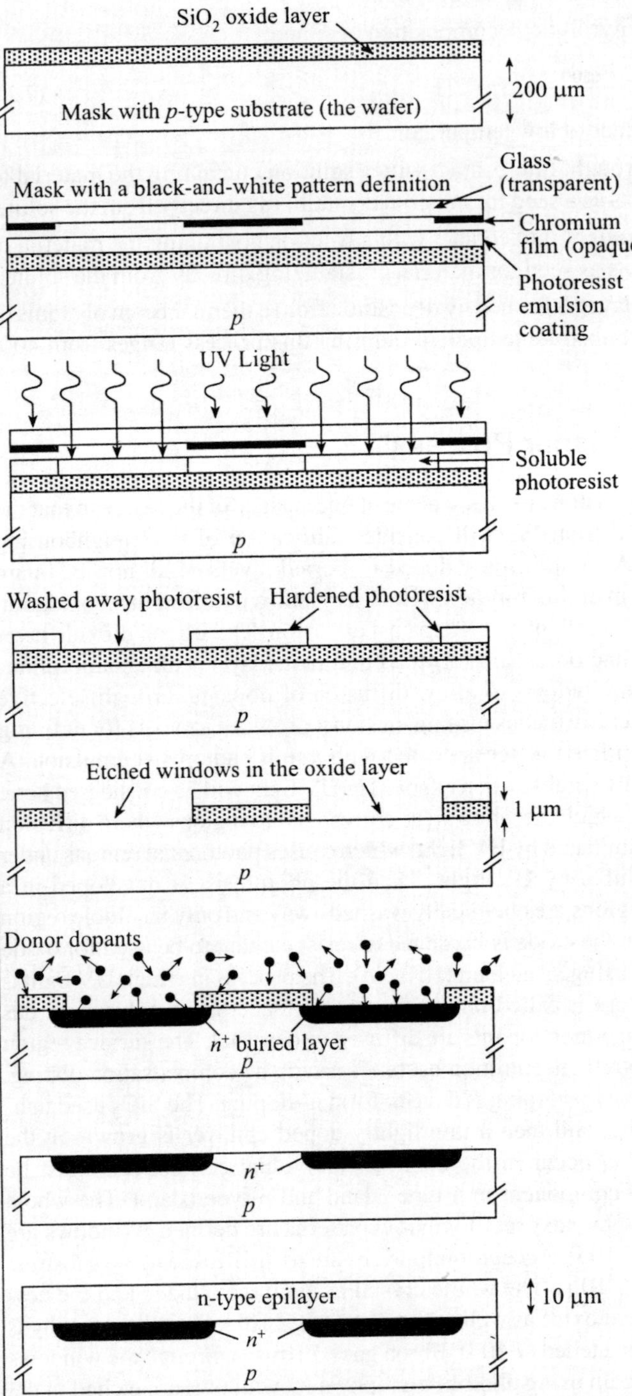

(a) Oxidation: Grow a thin oxide layer on the substrate surface in an oxidation furnace.

(b) Coat the surface of oxide with a thin photoresist film. Prepare a glass mask with a transparent and opaque pattern representing a "positive image". Transparent areas correspond to windows.

(c) Illuminate with UV radiation. Illuminated resist areas become soluble.

(d) Develop photoresist. This leaves only the regions of hardened resist.

(e) Etch windows in the oxide to allow dopant diffusion into the wafer.

(f) Diffuse donor dopants into the wafer in a diffusion furnace. Oxide serves as a mask against diffusion

(g) Etch away all the oxide from the surface.

(h) Grow an epilayer on the surface by vapor phase epitaxy in a reactor.

Fig. 7.10(a-h) Planar Process of IC fabrication

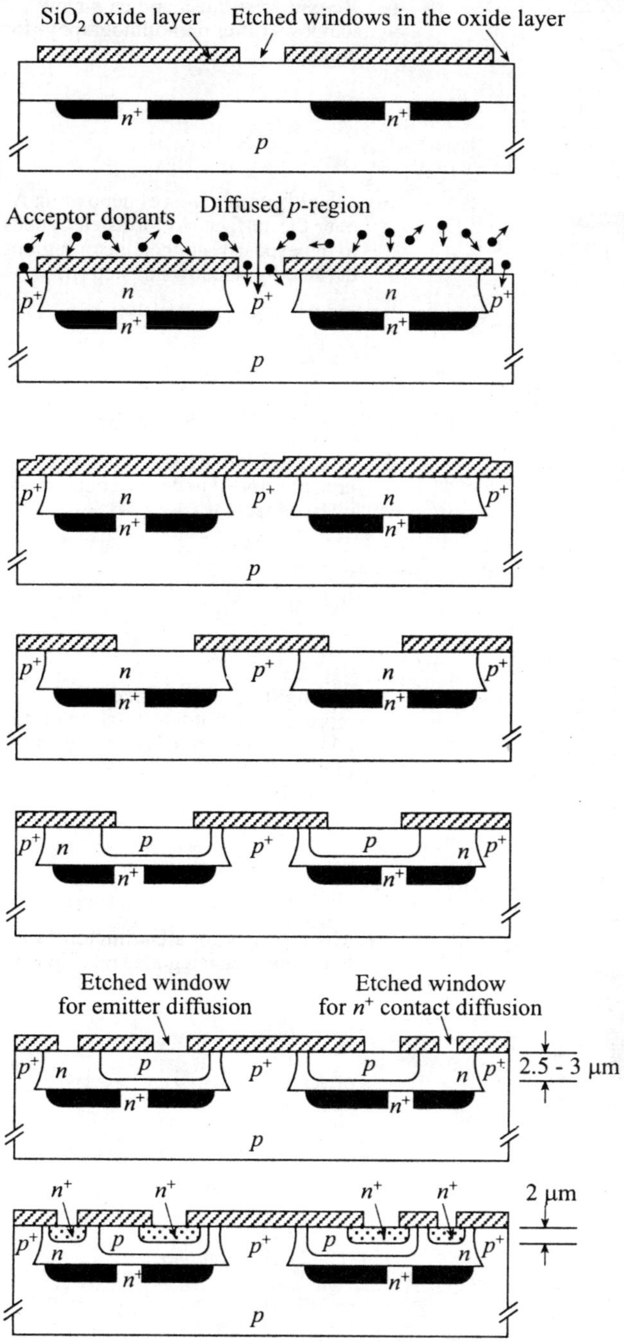

(i) Grow an oxide over the whole surface and then etch windows for isolation diffusion (after photolithography).

(j) Diffuse acceptor dopants through the windows in the oxide.

(k) Reoxidize the whole wafer surface.

(l) Use photolithography and etching to open windows in the oxide for base diffusion.

(m) Diffuse acceptor dopants through the windows to form the base region.

(n) Reoxidize surface and etch new windows (after photolithography) for n^+ diffusion for the emitter region and collector contact.

(o) Diffuse donor dopants through the windows for the emitter and the n^+ ohmic contacts to collector.

Fig. 7.10(i-o) Planar Process of IC fabrication

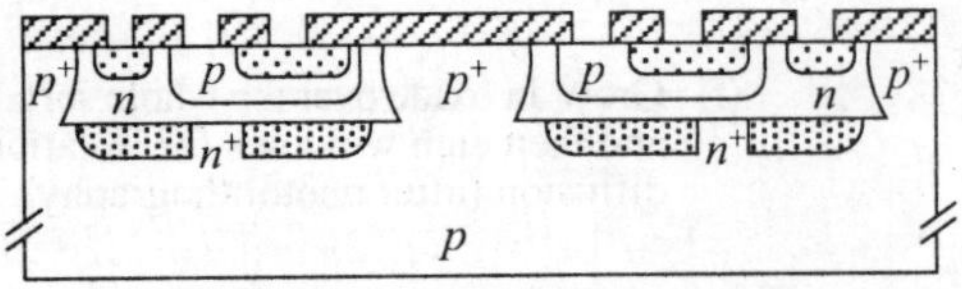

(p) Reoxidize surface, and etch new windows (after photolithography) for contacts.

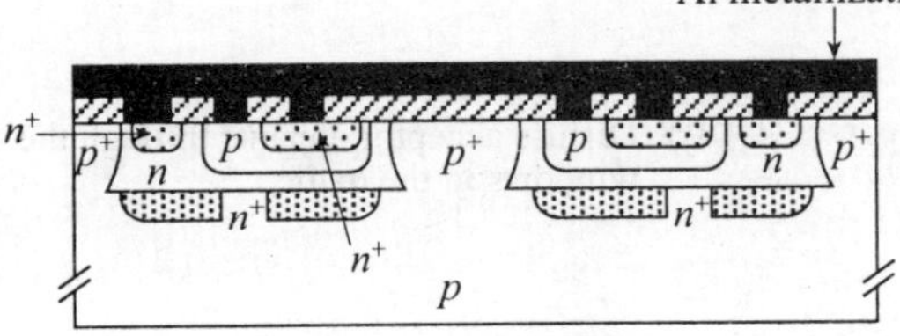

(q) Metallization involves depositing Al over the surface. Al forms electrodes to the exposed semiconductor regions defined by windows in step (p).

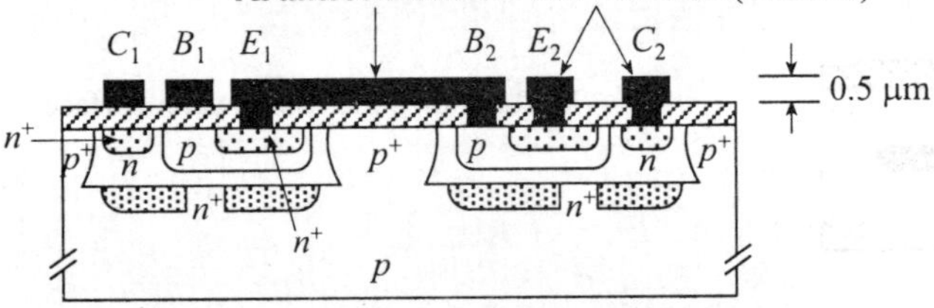

(r) Using a suitable mask, etch away unnecessary Al to define contacts and interconnects.

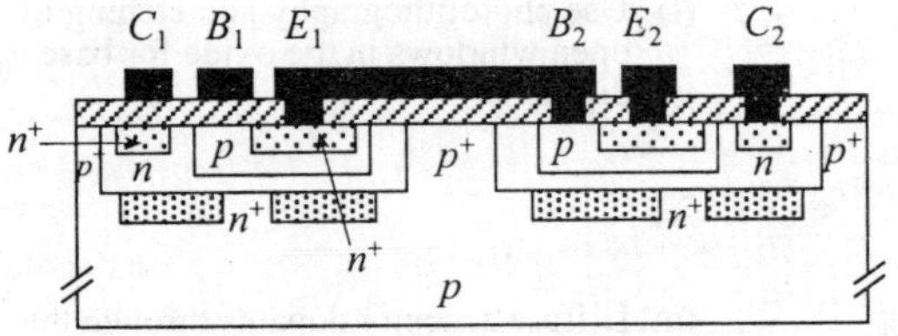

(s) A simplified and idealized representation of the structure in step (r) without the curved boundary effects of diffusion.

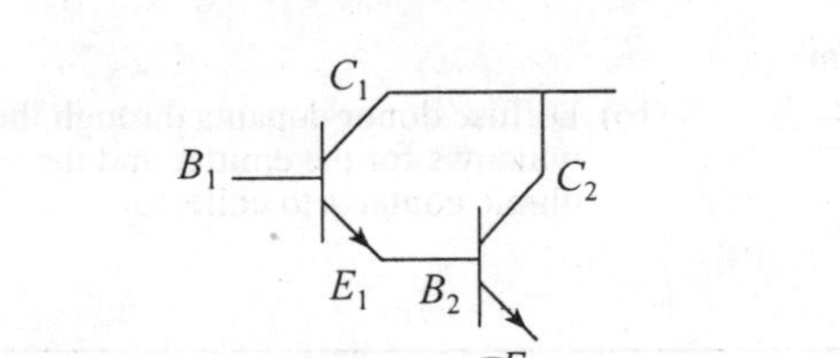

(t) The top view for a Darlington pair – two connected integrated npn bipolars.

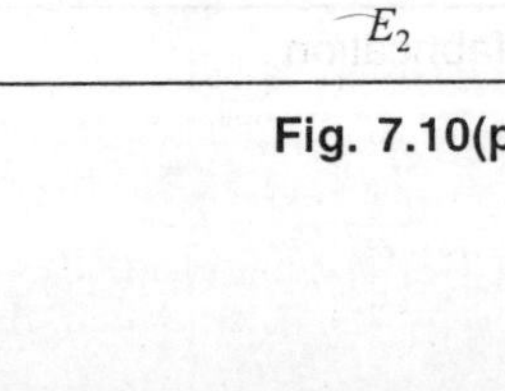

(u) The Darlington pair.

Fig. 7.10(p-u) Planar Process of IC fabrication

reoxydized and using photolithography, the oxide is masked by a photoresist pattern and windows are etched where contacts are needed (fig 7.10p). Aluminum is coated over the whole surface of the wafer (7.10q). The aluminum covering the whole surface area is etched selectively to remove all unwanted aluminum except where the metal is needed for interconnection (7.10r). After the formation of interconnection, each chip on the offer is electrically tested and those failing would be marked and discarded. After the testing, the wafer is diced, scribed with a diamond stylus, to individual IC chips. Each chip is mounted, bonded to leads, and encapsulated in a package that is suitable for application.

7.5 Metallization, Contacts

The requirement of low-resistivity materials leads one to consider the use of metals as interconnections. Aluminum is mostly used being cheaper and adheres to SiO_2. Aluminum may contact semiconductor regions in different ways as shown in fig 7.11. Aluminum contact to p-type Si normally results in a good ohmic contact for doping levels exceeding 10^{16} cm^{-3}. But for lightly doped n-region, Schottky junction can be formed between metal and semiconductor.

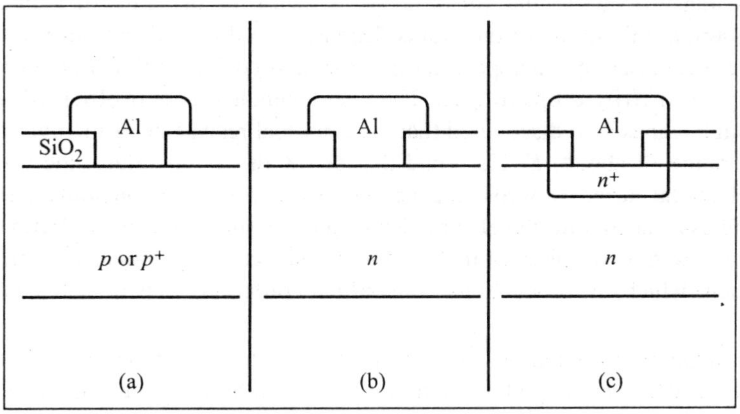

Fig. 7.11 Aluminum to p-type sc forms ohmic contact (a), aluminum to n-type forms Schottky junction (b) and Aluminum to n$^+$ region forms practical non-linear ohmic contact (c).

There is small resistance associated with an ohmic contact between two materials.

$$R_c = \frac{\rho_c}{A} \tag{7.25}$$

where ρ_c – specific contact resistivity, ohm-m^2,

A – area of the contact.

Aluminum contacts to silicides suffer from pitting and spiking problems. To circumvent these problems, an intermediate layer of metal is used to prevent silicon diffusion. In fig 7.11, titanium-tungsten (TiW) is used as a barrier metal over silicides in the contact regions of both bipolar and MOS technologies. The final contact consists of a sandwich of a silicide over the diffusion, followed by TiW diffusion barrier, and completed with aluminum copper interconnection metallization.

In most of the applications, a single layer of metal does not provide sufficient capability to fully interconnect complex VLSI chips. So, many processes use two or three levels of polysilicon as well as several levels of metallization in order to ensure wirability and provide adequate power distribution.

Superconductivity

8

Chapter

8.1 Introduction

The electrical resistance of metal and alloys decreases with the decrease of temperature. At lower temperature, the thermal vibrations of the atoms decrease. So due to decrease in randomness of free electrons, resistance decreases. When a specimen of material is cooled sufficiently low temperature in the liquid helium range, resistivity drops to zero. The phenomenon was first observed in 1911 by Dutch Physicist Kamerlingh Onnes who discovered that the electrical resistance of mercury drops to zero (see Fig 8.1.) at temperature of about $4K$. He found that below this temperature, the resistance remained effectively zero. Electrical current flowed in the mercury in the absence of potential difference and without causing the heating associated with the electrical resistance in the conductors. Onnes declared that "no doubt was left of a new state of mercury in which its resistance has practically vanished. Mercury has passed into a new state which on account of its extra ordinary properties may be called the superconductive state"

When a superconductor is cooled below a certain temperature, there is not only an abrupt loss of electrical resistance as discovered by Onnes but there are also abrupt changes in many properties of the materials. All these changes take place at a same temperature which is known as critical temperature (Tc). Below the critical temperature, the material is **superconductive state** and above the critical temperature, the material is *normal state*. The temperature at which transition between the superconductive state and normal state occurs is in the absence of magnetic field independent of shape or size of the specimen. It is a characteristic of material.

Before 1986, the superconductivity was rather low temperature phenomenon and is a definite obstacle to its technological exploitation, since it means that a superconducting device has to be accompanied refrigeration equipment of some kind to keep the temperature sufficiently low. Even only $125K$ or just one decade above the temperature, i.e. critical temperature has obtained in the compound $Tl_2Ba_2Ca_2Cu_3O_{16}$ (TBCCO) in the IBM institute. Most of the scientists of the world are working to obtain at room temperature. Experimentally and theoretically they have concluded that:

(I) No ferromagnetic or anti-ferromagnetic materials are superconductor. There is diamagnetic effect in superconductor.
(II) In metallic elements, superconductivity is only found when the number of valence electrons per atom lies between two and eight.
(III) In transition metal elements, the critical temperature varies in a periodic manner with the number of valence electrons per atom and exhibits sharp maximum. In non-transition elements, the critical temperature increases as the number of valence electrons per atom increases.

(IV) Low carrier concentration groups of superconductors are obtained by the addition of large amounts of appropriate impurities to semiconductors. Such as Ge, Tl etc.

The superconductivity, over these years has flourished as a field of scientific and technological endeavor, leading to the produce of many Nobel Laureates in physics. The Technological approach of superconductor is increasing rapidly, particularity in relation to computer switching element, bolometers, electromagnetic transportation, experimental science etc.

8.2 Historical Background

The history towards the direction of superconductivity starts from the liquefaction of helium by *Kamerlingh Onnes* in 1908. In 1911 *Onnes* discovered superconductivity in Hg at 4.2 *K*. In 1913, the superconductivity of lead at 72 *K* was discovered. It was observed in 1914 that superconductivity is destroyed by magnetic field. At critical magnetic field the superconductive state is destroyed. This discovery leads two conclusions of obstacle in the superconducting state i.e. temperature barrier and magnetic (or, corresponding critical current) barrier.

The magnetic or current barrier was overcome by the discovery of superconducting state in alloys in 1930s. Such types of superconductors are called type II superconductor. For such materials, superconducting state vanishes only above very high magnetic field of critical value Hc_2 (*T*). Such as for Nb_3Sn, it is about 200 *K Oersted*. The endeavors to overcome the temperature continued until the investigation of metal oxide ceramics $BaPb_{1-x}Bi_xO_3$ (BPBO) in 1974. In this system for $x = 0.25$, the *Tc* was 13*K*. Until 1986, the *Tc* remained 24*K* only for Nb_3Ge a record. But after that *Bednortz* and *Muller* detected the appearance of superconductivity in the ceramic La-Ba-Cu-O(LBCO) with *Tc* = 30-40*K*. Table 8.1. gives some data on superconductors.

Later, towards the end of 1986, the existence of high temperature superconductor (HTSC), which can be called **Neon superconductors,** was reliably proved in *Switzerland*, *Japan* and *USA* for La-Ba-Cu-O and La-Sr-Cu-O (LSCO). In the beginning of 1987, La^{57} was replaced by Y^{39} and this led to the appearance of *Nitrogen Superconductor* with $Tc \approx 80\text{-}100K$. For example in $(Y_{0.6}\,Ba_{0.4})_2\,CuO_{4-y}$ the $Tc = 93K$ at atmospheric pressure. Similar result were obtained soon afterwards in USSR and able to get up to 102*K* using *Yttrium* of certain proportions. In 1993, H.R. Otto and his group in Germany broke the record of the high temperature superconductivity of that decade getting 133*K* in Hg-Ba-Ca-Cu-O. Next year in 1994 Chinese scientists observed 164*K* at 30GPA pressure in the same compound. However, no stable transition temperature greater than 138*K* at normal pressure is observed to date. It was recored already in 1995 in the compound Hg-Tl-Ba-Ca-Cu-O. But now various scientists have obtained without a shadow of doubt, superconducting metal oxide ceramic of various compositions. Even for new Nitrogen superconductor, some non-reproducible results were obtained which could indicate the appearance of superconductivity with $Tc \approx 200\text{-}250K$. It can no means be ruled out that reliable information on the existence of room temperature superconductivity with $Tc \approx 250\text{-}300K$ will soon appear.

Theoretical study provides the causes of occurrence of superconducting state. There is really no good microscopic theory of superconductivity. What prevents from creating a new theory? Do scientists lack of additional facts or information or are waiting for the advent of a genius, a **New *Einstein***?

To get theoretical knowledge of superconducting it had to wait until the discovery of super fluidity of He - II in 1938. It become then more or less clear that superconductivity is the super fluidity of electrons in metal. Then due to *Bose - Einstein* condensation and super fluidity, it is concluded that electrons in metals can be converted into bounded electron pairs with zero total spin. How electrons usually repel, can form a

Table 8.1 Properties of some selected superconductors in chronological order.

Year	T_C(K)	*Material*	*Class*	*Crystal Structure*	*Type*	H_C* (MAm^{-1})
1911	4.2	*Hg*	Metal	Tetragonal	I	0.033
1913	6.2	*Pb*	Metal	*fcc*	I	0.064
1930	9.25	*Nb*	Metal	*bcc*	II	0.164
1940	15	*NbN*	Interstitial Compound	*NaCl*	II	12.2
1950	17	V_3Si	Intermetallic Compound	β-tungsten (W_3O)	II	12.4
1954	18	Nb_3Sn	Intermetallic Compound	W_3O	II	18.5
1960	10	*Nb-Ti*	Alloy	*bcc*	II	11.9
1964	0.7	$SrTiO_3$	Ceramic	Perovskite	II	small
1970	20.7	$Nb_3(Al, Ge)$	Intermetallic Compound	W_3O	II	34.0
1977	23	Nb_3Ge	Intermetallic Compound	W_3O	II	29.6
1986	34	$La_{1.85}Ba_{0.15}CuO_4$	Ceramic	Tetragonal	II	43
1987	90	$Yba_2Cu_3O_7$	Ceramic	Orthorhombic	II	111
1988	108	*Bi* cuprates	Ceramic	Orthorhombic	II	–
1988	125	*Tl* cuprates	Ceramic	Orthorhombic	II	–
1993	133	*HgBaCaCuO*	Ceramic	Orthorhombic	II	–
1994	164	*HgBaCaCuO at 30GPa*	Ceramic	Orthorhombic	II	–
1995	138	*HgTlBaCaCuO*	Ceramic	Orthorhombic	II	–
2001	39	MgB_2	Metallic		II	–

* *extrapolated to 0K.* H_{C2} *for type II materials.* (1 *Oersted* = 79.6 Am^{-1})

bound state? In 1954, *Ogy*, reported that electrons are bound in pairs and *Bose-Einstein* condensation takes place by experimental way. Later *Ginsberg* and *Landau* gave a theory i.e. ψ- theory. *Schafroth* put forth the idea of pairing in most clear way and unmitigated suggesting that even the electron-phonon interaction could be responsible for pairing, but left unanswered about the nature and mechanism of pairing.

In 1956 *Cooper* introduced first pairing of electron called *Cooper pair*. Then the next year in 1957, the microscopic theory called reliable theory was first constructed, only 46 years after the discovery of superconductivity. It is known as *Bardeen, Cooper* and *Schrieffer* (BCS) theory. This theory concludes the following important results:

- The critical temperature of superconductor is *Debye* temperature (θ_D) dependant i.e.

  $$Tc = 1.14\, \theta_D\, e^{-\frac{1}{\lambda}}$$ λ is coupling constant.

- The energy gap in superconductor at transition temperature is $Eg = 3.52\, K_B\, Tc$
- The specific heat capacity at transition temperature is $Ce = 1.43\, \gamma Tc$ when γ is the coefficient of electronic specific heat
- Tc depends on mass and observed as $Tc \propto M^{-\frac{1}{2}}$ called isotopes effect.

8.3 Electric and Thermodynamic Properties

The different empirical electrical and thermodynamic properties of superconductor will be presented.

DC Resistivity

The first change to be discovered in superconductivity was the vanishing of electrical resistance as mentioned in earlier section. At the critical temperature, the DC resistivity vanishes and below it remains zero as shown in Fig. 8.1. The careful investigations have shown that the resistivity of a metal in the super conducting state drops to less than one part in 10^{17} of its value in normal state.

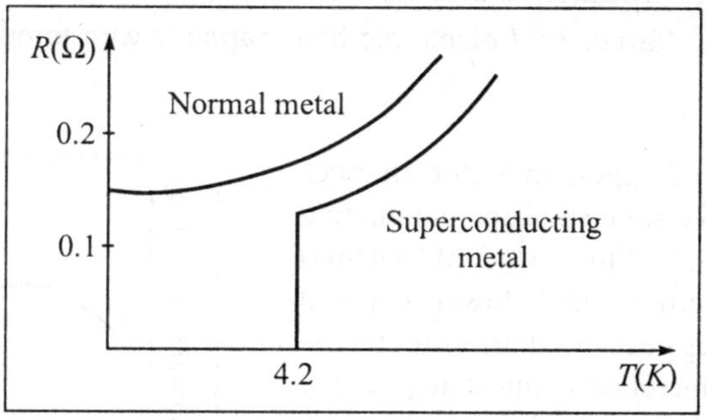

Fig. 8.1 Temperature dependence of the resistance of a normal metal and superconducting metal like Hg

AC Resistance

The zero resistance of the superconducting state is true only for DC of constant value. If the current is changing then a non zero resistance exists. For AC of a certain frequency, the resistance is both frequency as well as temperature dependent. It is also found that at high frequency the resistances in superconducting and normal states are equal. It is because, superconducting electrons are in a lower energy state than normal electrons, but if the frequency of the applied field is high enough, the photons of electromagnetic field have enough energy to exits superconducting electrons into the higher state where they behave as normal electrons. Such behavior is observed only above 10^{11} Hz.

Specific heat capacity

The specific heat capacity of a normal metal is in the form

$$C(T) = \gamma T + \beta T^3 \tag{8.1}$$

When first term represents the specific heat of electron in metal and the second is the contribution of lattice vibrations at low temperature. In superconducting state only electronic part is effected. So on subtracting the lattice contribution we will get as the form

$$C\mathrm{el}^{(T)} = B\, e^{-\frac{\phi}{\mathrm{KBT}}} \tag{8.2}$$

Which shows the experimental variation of the electronic specific heat capacity with temperature and existence of a finite gap (ϕ) in the energy spectrum of electron separating two states.

At critical temperature, the specific heat capacity of the material increases abruptly. Fig 8.2 shows the variation of electronic specific heat with temperature.

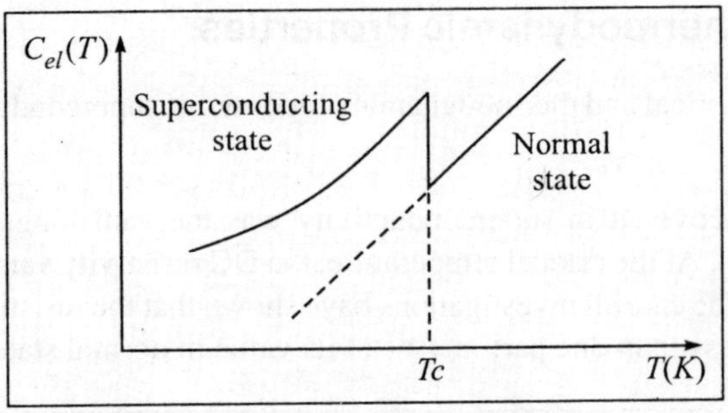

Fig. 8.2 Variation of electronic heat capacity with temperature

Thermal conductivity

The thermal conductivity of superconductor changes continuously between normal state and superconducting state as shown in Fig. 8.3. The value of thermal conductivity in superconducting state is lower than that in normal state which suggests that the electronic contribution drops in the superconducting state possibly playing no part in heat transfer. The thermal conductivity of *tin* at $2K$ is 34 wcm^{-1} K^{-1} for its normal state and 16 wcm$^{-1}K^{-1}$ for its supper conducting phase.

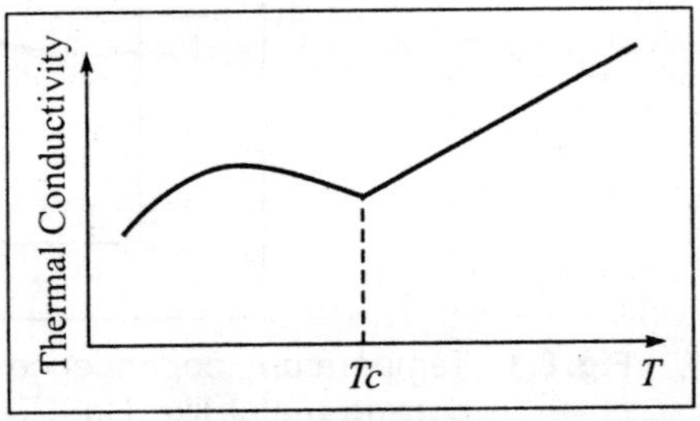

Fig. 8.3 Variation of thermal conductivity with temperature

The Energy Gap

It is well known that the exponential form is compatible with the thermal excitation across a gap in energy. This fact concludes an energy gap in the superconducting electron levels because of experimental nature of specific heat capacity in superconducting state of materials. Fundamentally, this gap differs from the energy gap in semiconductors or insulators. In superconductor the gap is related to Fermi gas and in semiconductor, the gap is attached to the lattice. The gap presents the flow of current in semiconductor whereas in superconductor current flows despite the presence of gap. From optical and many other methods, it is concluded that the gap energy is

$$E_g = 2\phi = 2\,\mathrm{b}\,K_B\,Tc \tag{8.3}$$

2b is constant and is about 3.5 for all materials. For gallium

$$E_g \cong 10^{-4}\,\mathrm{eV}$$

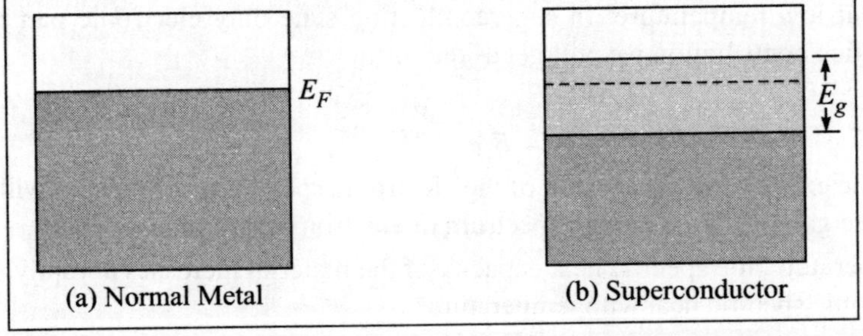

Fig. 8.4 Occupied and unoccupied sates in (a) normal metal and (b) superconductor

8.4 Magnetic Properties

The magnetic properties of superconductors are unusual and unexpected. But magnetic properties are fundamental to superconducting state. The different magnetic properties are reviewed below.

The Meissner Effect

If a superconductor is cooled in a magnetic field from above the transition temperature then at the transition temperature and below, the lines of induction of B are pushed out.

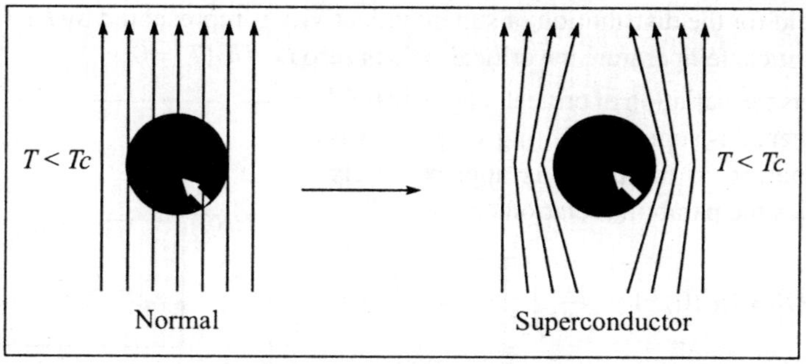

Fig. 8.5 Perfect diamagnetic effect in a superconducting state

This phenomenon was first discovered by *Meissner* and *Ochsenfeld* in 1933, and is called *Meissner* effect. It shows that a bulk superconductor behaves in an applied external field as it inside the specimen $B = 0$, i.e. perfect diamagnetic effect. Inside the supeconductor $B = 0$. Thus

$$B = \mu_0 (H + M) = 0$$

i.e.

$$H = -M \text{ and } \chi = \frac{M}{H} = -1$$

which is true for a perfect diamagnet.

The perfect diamagnetic behaviour of superconductor cannot be explained simply by considering zero resistivity. Now let an electric field E would accelerate superconducting electrons then

$$m\left(\frac{dv}{dt}\right) = eE \text{ and super current density}$$

$$J_s = n_s e v$$

$$\left(\frac{\partial Js}{\partial t}\right) = \left(\frac{n_s e^2}{m}\right) E \tag{8.4}$$

i.e. if there is zero resistance, relaxational time would accelerate infinitely, which is not possible. So E must be zero inside the metal. From *Maxwell*'s equation.

$$\nabla \times E = -\frac{\partial B}{\partial t} = 0$$

i.e.

$$B = \text{Constant} \tag{8.5}$$

Thus B is constant in perfect conductor, which is contradiction to the *Meissner* effect. The behaviour of superconductor is different from that of a perfect conductor. Hence the superconducting state has mutually independent properties i.e. zero resistivity and perfect diamagnetism i.e.

$$E = 0 \text{ and } B = 0 \tag{8.6}$$

Materials showing such behaviour are known as soft or type I superconductors.

Critical Magnetic Field

A sufficiently strong magnetic field will destroy superconductor. The threshold or critical value of the applied magnetic field for the distribution of superconductivity is represented by $Hc\ (T)$ and is a function of temperature. At critical temperature the critical field is zero i.e. $Hc\ (T) = 0$

The Fig. 8.6 depicts the variation of critical magnetic field with temperature where S is superconducting state and N is normal state. The nature of the curve is approximately parabolic and satisfies the parabolic relation

$$Hc\ (T) = Hc\ (0) \left[1 - \left(\frac{T}{Tc}\right)^2\right] \tag{8.7}$$

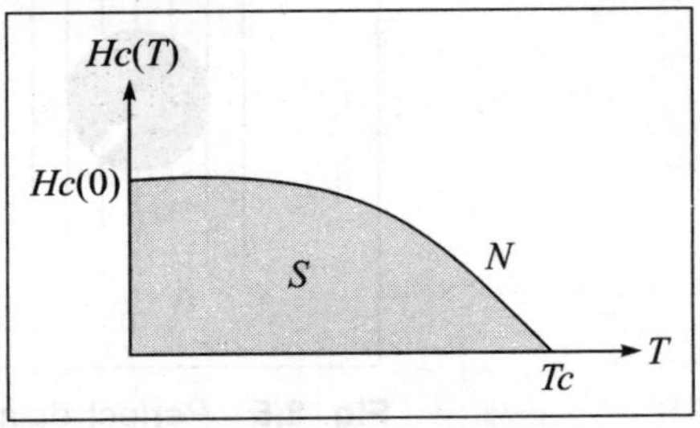

Fig. 8.6 Variation of critical field with temperature

where $Hc\ (0)$ is the critical field at $0K$. Thus we find that the superconducting state is stable only in some definite ranges of magnetic fields and temperatures. For higher fields and temperatures, normal state is more stable.

Critical Current

In addition to $Hc\ (T)$ and Tc, a third critical parameter is often specified to a conductor. A given superconductor cannot support an electric current density greater than a critical value Jc which is temperature dependent and reaches a maximum at $T = 0K$. The critical current density and critical magnetic field are closely related because on passing current necessarily produces a magnetic field.

If a wire of radius a of type I superconductor carries a current I, a surface magnetic field H_I is set up. In addition if a transverse magnetic field Ha is applied to the wire, then the condition for transition to normal state is

$$Hc = H_I + 2Ha$$

i.e. $$H_1 = Hc - 2Ha$$

where $H_I = \dfrac{1}{2\pi a}$ and at critical value $I = Ic$

$$\therefore \quad \frac{1c}{2\pi a} = Hc - 2Ha$$

i.e. $$Ic = 2\pi a\,(Hc - 2Ha) \tag{8.8}$$

Hence the critical current decreases linearly with increase of the applied field until $Ic = 0$ at $Ha = \dfrac{Hc}{2}$

This is also called *Silsbee's* rule. Fig 8.7 shows the variation of critical current density in a wire as a function of transverse applied field for pure and impure materials.

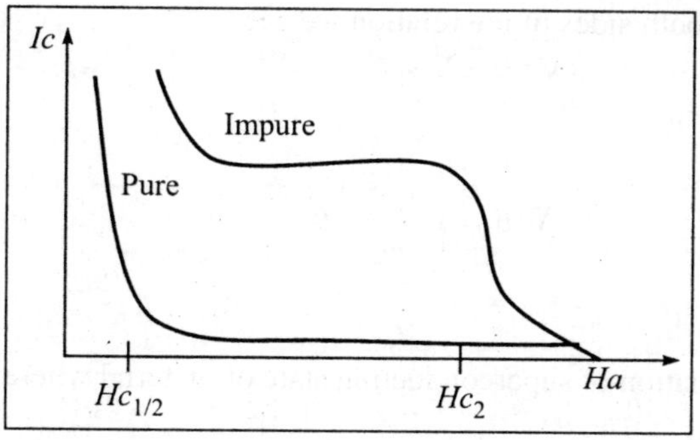

Fig. 8.7 Variation of critical current of pure and impure materials with transverse applied field

The London Equation

Two brothers *F.* and *H. London* in 1935 explained *Meissner* effect with two fluid models based on super fluid helium.

Electrical condition in normal state of a metal is described by *Ohm's* law which is to be modified to describe condition and *Miessner* effect in superconducting state. When a transitory electric field used to induce current in a sample, only the superconducting electrons flow while the normal electrons remained stationary and can be ignored as in equation (8.4)

i.e.
$$\frac{dJs}{dt} = \left(\frac{n_s e^2}{m}\right) E$$

where n_s is the density of superconducting electrons.

Taking curl of both sides of this equation we get,

$$\frac{d}{dt}(\nabla \times Js) = \left(\frac{n_s e^2}{m}\right) \nabla \times E$$

using *Maxwell's* equation $\nabla \times E = -\frac{\partial B}{\partial t}$ we get

$$\frac{\partial}{\partial t}\left[\nabla \times Js + \left(\frac{n_s e^2}{m}\right) B\right] = 0$$

This equation cannot explain *Meissener* effect unless restricting and requiring the expression in square bracket be zero not just time dependence. So

$$\nabla \times Js = -\left(\frac{n_s e^2}{m}\right) B \tag{8.9}$$

Again we have from *Maxwell's* equation

$$\nabla \times B = Js$$

Again taking the curl on both sides of the relation we get

$$-\nabla^2 B = \nabla \times Js$$

Hence from (8.9) we get

$$\nabla^2 B = \left(\frac{n_s e^2}{m}\right) B \qquad (8.10)$$

This is called *London* equation of superconducting state of material where $\left(\frac{m}{n_s e^2}\right)^{\frac{1}{2}}$ is called *London* penetration depth and is represented by λ_L. The *London* equation is second order differential equation in B where ∇^2 is *Laplace* operator. Now the equation is

$$\nabla^2 B = \frac{B}{\lambda_L^2}$$

This equation occurs for the *Meissner* effect because it does not allow a solution uniform in space i.e. uniform magnetic field cannot exist in a superconductor. In pure superconducting state, the only field allowed is exponentially decayed as we go in from an external surface. Let a semi-infinite superconductor occupies the space on the positive side of x-axis, then the solution of the *London* equation for this case will be

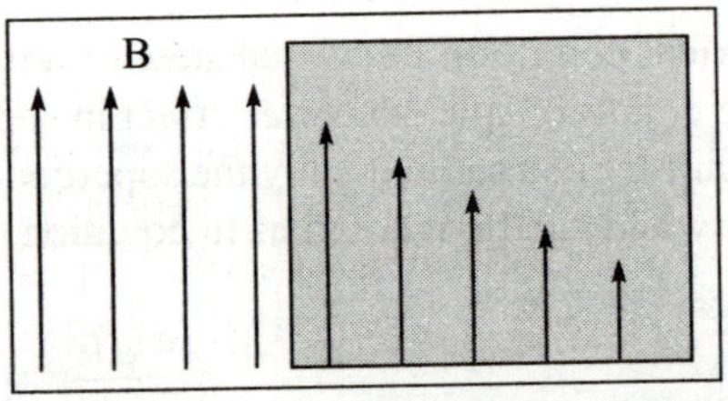

Fig. 8.8 Penetration of an applied magnetic field into a semi-infinite superconductor.

$$B(x) = B(0)\, e^{-\frac{x}{\lambda_L}} \qquad (8.11)$$

where $B(0)$ is the field at the plane boundary. The penetration depth λ_L measures the depth of penetration in which field decreases by a factor e^{-1}. If thickness of the specimen is very less than λ_L, *Meissner* effect is incomplete and it $\lambda_L << x$, field damps out exponentially inside the superconductor. So λ_L is the fundamental length that characterises a superconductor.

Coherence Length

The superconducting is due to mutual interaction and correlation of the behaviour of electrons which extend over a considerable distance. The maximum distance up to which the states of pair electrons (cooper pairs) are correlated to produce superconducting state is called coherence length. Hence coherence length is the measure of the minimum spatial extent of a transition layer between normal and superconductor. Any spatial variation in the state of an electronic system requires extra kinetic energy. A modulation of an eigen function increases the kinetic energy because of the increase of integral of $\frac{d\phi}{dx^2}$.

Let the kinetic energy of a plane wave, $\psi = e^{ikx}$ is given by

$$E = \int \psi \left(-\frac{h^2}{2m}\nabla^2\right) \psi^* \, dx = -\frac{h^2}{2m}\int \psi^* \left(\frac{\partial^2 \psi}{\partial x^2}\right) dx$$

i.e.
$$E = \frac{h^2k^2}{2m} \tag{8.12}$$

where $\psi\psi^* = 1$ but for the strongly modulated wave function

$\phi(x) = \frac{1}{\sqrt{2}}(e^{i(k+q)x} + e^{ikx})$ the corresponding kinetic energy is

$$E = \int \phi^* \left(-\frac{h^2}{2m}\nabla^2\right)\phi\, dx$$

$$= -\frac{h^2}{2m}\int \phi^* \frac{\partial^2\phi}{\partial x^2}dx$$

$$= \frac{1}{2}\left(\frac{h^2}{2m}\right)[(k+q)^2 + k^2]$$

$$E \approx \frac{h^2}{2m}k^2 + \frac{h^2}{2m}kq \tag{8.13}$$

where $q << k$. Thus the increase of energy required to modulate the wave is $\frac{h^2kq}{2m}$. If this increase exceeds the energy gap Eg, superconducting state will be destroyed. The critical value of q is q_0 of the modulation wave vector for normal $k = k_F$ is then

$$Eg = \frac{h^2}{2m}k_F\, q_0 \tag{8.14}$$

Now we can define the intrinsic coherence length ξ_0 as

$$\xi_0 = \frac{1}{q_0} = \frac{h^2k_F}{2mEg} = \frac{hV_F}{2Eg} \tag{8.15}$$

where V_F is electron velocity in *Fermi* surface. Hence coherence length is a measure of the distance within which the gap parameter (Eg) cannot change drastically in a spatially varying magnetic field.

8.5 Type *I* and Type *II* Superconductor

The superconductor in which the *Meissner* effect is complete up to critical field Hc is called type *I* superconductor. The magnetic behaviour of a typical type *I* superconducting is shown in Fig. 8.9. These material give away their superconductivity at lower field strength so these are referred to as soft superconductors. Pure specimens of various metals exhibits this type of behaviour.

Surface energy of a material is a result of magnetic field which is positive for type *I* superconductor. Another important feature of the type *I* superconductor is exhibiting an intermediate state between superconducting state and normal state. However a cylinder placed in parallel magnetic field does not exhibit intermediate state. The coherence length of type *I* superconductor is greater than penetration depth. Such as for *Sn*, $\xi_0 = 23 \times 10^{-6}$ cm and $\lambda_L = 3.4 \times 10^{-6}$ cm and for *Al* $\xi_0 = 160 \times 10^{-6}$ cm and $\lambda_L = 1.6 \times 10^{-6}$ cm. The mean free path of type *I* superconductor is greater than penetration depth. Pure lead having $\xi_0 = 8.3 \times 10^{-6}$ cm and $\lambda_L = 3.7 \times 10^{-6}$ is a type *I* superconductor exhibiting the magnetizing curve 1 in Fig. 8.10.

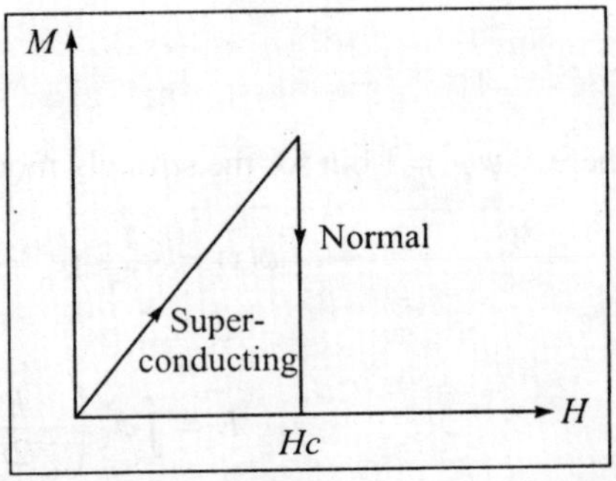

Fig. 8.9 Magnetization curve of Type *I*

But when some amount of indium is doped to the lead, the behaviour will no longer type *I*. It shows the curve 2 type behaviour, which is characteristics of type *II* superconductor. These types of materials tend to be alloys, ceramic or hard transition metals with high value of electrical resistivity at normal state i.e. low value of mean free path in normal state. In such type of materials, superconducting state exists up to Hc_2. Between the lower critical field Hc_1 and upper critical field Hc_2, $B \neq 0$ i.e. *Meissner* effect is incomplete. The region between Hc_1 and Hc_2 is called **Vortex State**. The critical field of He is 4.8×10^4 A/m at 4*K* and when 20% indium is added it becomes type *II* superconducing of which $Hc_1 = 3.6 \times 10^3$ A/m and $Hc_2 = 2.9 \times 10^5$ A/m. Surface energy of type *II* superconductor is negative so vortex state or mixed state is stable. It is also observed that the penetration depth (λ_L) of these superconductors is larger than coherence length (ξ_0). These superconductors are also called hard or impure superconductor.

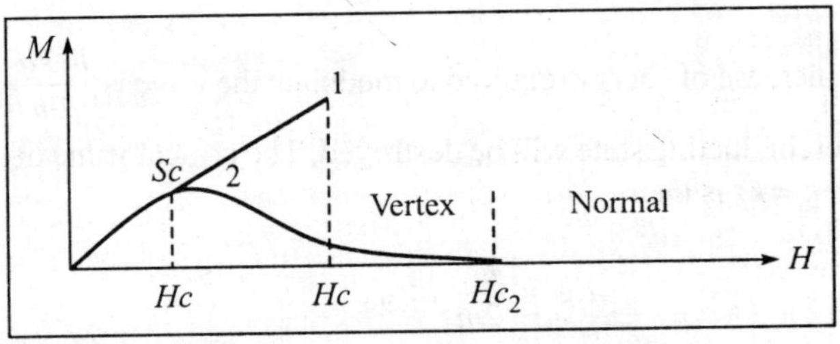

Fig. 8.10 Magnetization Curve

8.6 The BCS Theory

In 1957 almost half a century after the discovery of superconductivity, a comprehensive microscopic theory of superconducting phenomenon was proposed by *Bardeen*, *Cooper* and *Schrieffer* called BCS theory. This theory helps to overcome the difficulties to pair that the electrons are fermions and repulsive force among them take away from each other. The main BCS theory is described using the basic three phenomena:

Electron-phonon interaction

According to *Frohlich*, when an electron moves through a crystal it produce lattice distortion and sets the heavier ions into slow vibrations. Since the electron moves very fast leaving this region much before the oscillation can die off. This slightly increases the density of the positive ion cores in the intermediate region of the electron and polarizes the crystal. Other electrons in this region with high density of positive ions are attracted. Hence the electrons are attracted to each other via the movement of ions.

The interactions between (I) electron-electron and (II) electron-phonon or lattice will be strong enough to overcome the coulomb force. The net effect is the attraction of two electrons via a lattice distortion to form a pair of electrons known as **cooper pair**. This can also be expressed by **isotope effect**. The large electron-phonon interaction would be expected to lead to high temperature superconductor and also leads high resistance in normal state.

Cooper pair

As described above the lattice vibration plays an important role and due to interaction if wave vector (k_1) of an electron is equal and opposite to another ($-k_2$), results an attractive force with same energy, which causes the electrons to exit in pair. It is also shown that the interaction is larger for the electrons with opposite spins because they have lager probability of being close to each other. The coherence length is used to measure the size of cooper pair. If $B_E = 2\,Eg$ be the binding energy of the pairs, i.e. cooper pairs then

$$\xi_0 = \frac{hV_F}{B_E}$$

i.e. the size of pair depends on Fermi velocity and binding energy. It is of the order of 10^{-4} cm. All electrons in the metal tend to form such pairs k with $-k$ and that this electron gas undergoes a *Bose - Einstein* condensation.

BCS Ground State

In non-interacting Fermi gas, all state below Fermi- surface are occupied and above it are vacant. The lowest exited state is separated from ground state by an arbitrary low excitation energy. But the phonon assisted attractive interaction between electrons gives rise to the BCS ground state. This is separated from the lowest exited state by a finite energy gap *Eg*. The BCS state appears to have a higher kinetic energy then the Fermi state, but the attractive potential energy of BCS state decreases, i.e. the BCS state is more stable than Fermi state

The BCS ground state differs from ground state of non-interacting Fermi gas.

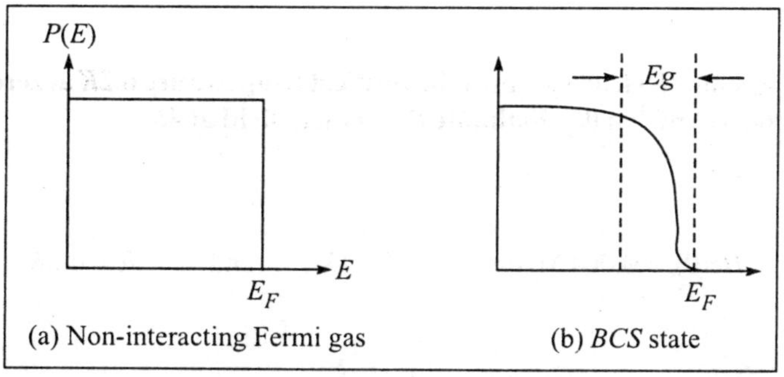

Fig. 8.11 Probability of occupancy versus energy of (a) non-interacting Fermi gas and (b) the BCS ground state

The superconducting state is a condensed in the source that a finite energy *Eg* is required to produce an exited state of the whole system. It is a bit like a semiconductor, but in semiconductor energy gap is tied to the Brillowin zone and the superconductor energy gap is carried by Fermi surface.

8.7 Material Aspects

First superconducting state was observed in *Hg* by *Kamerlingh Onnes* in 1911. The *Tc* of *Hg* was 4.2*K*. Later 1973 *J.R. Gavaler* observed 22.3*K* in Nb_3Ge. A new class of oxide ceramic superconductor having the critical temperature greater than 30*K* was discovered by *Bednorz* and *Muller* in 1986 which developed a new era in the field of superconductivity. These are called high *Tc* - superconductors. The first group of such materials are $La_{2-x} M_x Cu O_4$ (M = Ba, Sr, Ca) with *Tc* ranging from 35*K* to 40*K* and is referred to as 214 system. Another system of the material is $M Ba_2 Cu_3 O_{7-y}$ with $y = 0.2$ and (M = Te, Y, Nd, Sm etc) is called 123 system. Other system are Bi-Sr-Ca-Cu-O (BSCCO) with onset temperature of 110*K* and Tl-Ba-Ca-Cu-O (TBCCO) of 118*K* are also found.

The BSCCO and TBCCO contain planes of Cu- and O- atoms. But in YBCO compound have both planes and chains of Cu- and -O- atoms. Planes plays the major role in generating superconductivity while chains acts as electron reservoirs which can be filled or emptied either by changing the O- *Stoichimetry* or doping. These compounds have orthorhombic structure with distortion and almost always exhibit twinning. It was speculated that the twin planes in YBCO samples are a source of high temperature superconductivity. But this idea became less convincing after the development of *Tl* and *Bi* compound which do not have twin plane.

Properties of high T_c superconductor:

1. **High T_C** which is an appreciable function of *Debye* temperature (θ_D)and is given by $Tl \approx \theta_D e^{-\frac{1}{\lambda}}$ where λ is coupling constant characterizing the intensity of attraction.
2. **Short coherence length:** which is temperature dependence and is the consequence of the BCS theory $\xi = \xi_0 \left(1 - \frac{T}{Tc}\right)^{-\frac{1}{2}}$
3. **Large anisotropy:** *Tl* based superconductors are most an isotropic than any other superconductors. Such as Bi, La or Y based compounds.

8.8 Examples

1. The superconducting state of a lead specimen has critical temperature 6.2*K* at zero magnetic field and the critical field is 0.064 MAm^{-1} at 0*K*. Estimate the critical field at 4*K*.

Solution:

We have

$$Hc(0) = 0.064 \text{ MAm}^{-1}, \qquad T = 4K \qquad \text{and} \qquad Tc = 6.2K$$

Also we have,

$$Hc = Hc(0)\left[1 - \left(\frac{T}{Tc}\right)^2\right]$$

$$= 0.064\left[1 - \left(\frac{4}{6.2}\right)^2\right] = 0.037 \text{ MAm}^{-1}.$$

Hence the critical field of *Pb* at 4*k* is 0.037MAm^{-1}.

2. Calculate the size of a cooper pair if the gap energy is 10^{-3} eV and Fermi velocity of the electron is 10^6 ms^{-1}.

We have,

$$h = 1.055\times 10^{-34}\text{ Js} \qquad V_F = 10^6\text{ ms}^{-1} \qquad BE = 10^{-3}\text{eV} = 10^{-3}\times 1.6\times 10^{-19}\text{J}$$

Then using relation

$$\varepsilon_0 = \frac{hV_F}{B_E} = \frac{1.055\times 10^{-34}\times 10^6}{1.6\times 10^{-22}} = 4\times 10^{-7}\text{ m}$$

Thus the size of cooper pair is 4×10^{-7} m and is comparatively longer than the distance between two electrons ($\approx 10^{-10}$m)

Exercise

1. Determine the temperature at which the critical field becomes half of its value at $0k$ if the critical temperature of a superconductor at zero magnetic field is T_C. [$0.707T_C$]
2. The BCS theory of superconductivity shows that the superconducting energy gap at $0k$ is $3.5k_B T_C$. Calculate the maximum wavelength of microware radiation that will be absorbed at $0k$ in lead. [0.57 mm]
3. Determine the critical current which can flow through a long thin superconducting wire of aluminium of diameter 1mm. The critical magnetic field for aluminium is 8×10^3 Am^{-1}. [24.80 A]
4. Calculate the coherence length of an aluminium if the size of the energy gap is 3.4×10^{-4} eV and Fermi velocity is 2.02×10^6 ms^{-1}. [3.9×10^{-6}m]
5. For a superconducting specimen, the critical field are 1.4×10^5 and 4.2×10^5 Am^{-1} for $14k$ and $13k$. Estimate the transition temperature and critical fields at $0k$ and $4.2k$. [$14.5k$, 20.7×10^5Am^{-1} and 18.9×10^5Am^{-1}]
6. A niobium superconducting solenoid carries current 20A to produce 6×10^4Am^{-1}. Estimate the minimum diameter of wire that may be used in the solenoid if it is immersed into liquid helium at $4.2k$. The critical field for niobium is 1.56×10^5Am^{-1}. [1.16mm]

Einstein's Equation

A

Appendix

Electromagnetic waves travel through free space at a speed 3×10^8 ms^{-1} which is the highest value of speed of particles and mechanical waves. *Einstein* postulated in his well known relativity theory that speed of light has the same value for all observers, independent of the motion of the light source or observer. The relativity theory relates every physical quantity with velocity of light. The time dilation, length contraction, variation of mass with velocity, velocity addition, mass-energy equivalent etc. are main consequences of special theory of relativity.

The *Einstein*'s equation is the relation between mass and energy i.e. mass can be converted into energy. The equivalence factor is velocity of light. This equation is used not only in wave-mechanical study but also in nuclear science. To derive this relation, let us consider a particle of mass *m* acted upon by a force *F* in the same direction as its velocity *v* produces a displacement *dx*. Thus the increase in kinetic energy of the particle is represented by work done (*Fdx*). Hence

$$dw = Fdx = \frac{d}{dt}(mv)\,dx$$

$$= \left(m\frac{dv}{dt} + v\frac{dm}{dt}\right)dx$$

$$dw = mvdv + v^2 dm \qquad \left[\because v = \frac{dx}{dt}\right] \qquad A(1)$$

According to the variation of mass with velocity of the relativity theory,

$$m = \frac{m_0}{\sqrt{1 - \frac{v^2}{c^2}}}$$

or,

$$m^2 = \frac{m_0{}^2}{\left(1 - \frac{v^2}{c^2}\right)} \qquad A(2)$$

or,

$$m^2c^2 = m_0{}^2c^2 + m^2v^2$$

Differentiating it we get,

$$2mc^2dm = 0 + 2mdmv^2 + 2vdvm^2$$

or,

$$c^2\,dm = mv\,dv + v^2\,dm$$

Hence from equation $A(1)$ we have,

$$dw = c^2\,dm \text{ and integrating both sides}$$

$$w = c^2 \int_{m_0}^{m} dm = c^2(m - m_0) \qquad A(3)$$

This is the relativistic formula for kinetic energy. At rest $m = m_0$ from A (2), so m_0c^2 is the rest mass energy. Thus the total energy of the particle is

$$E = w + m_0c^2 = (mc^2 - m_0c^2) + m_0c^2$$

$$E = mc^2 \qquad A(4)$$

which is well known *Einstein*'s mass energy relation which states that mass can be converted into energy or vice versa.

Collection of Particles

Appendix B

In general, there are $3N$ equations of motion of second order for a system containing N-particles, which needs three-dimensional space to specify a particle. Such space is known as configurational space. For general consideration, it is more convenient to write $6N$ partial differential equations of the first order in place of $3N$ equations of second order. For this purpose, there must be $6N$ degree of freedoms or $6N$ dimensional space called **Phase Space**. In this space, a single particle is specified by 3-position co-ordinates and 3-momentum co-ordinates.

The physical quantities of interest in a thermodynamic system are usually the time average over the trajectory of particles in the phase space. In general, calculation of time average is a very complex task. There are a very large number of orders of 10^{23} of particles in a system which interact with one another in a complicated manner. The detailed kinematics of all these particles is a hopelessly tedious task even for a very small or simple system like an ideal gas. But *Gibbs* and *Boltzmann* found statistical method by which we can easily calculate the averages and resolving the impasse.

There are large numbers of points in phase space, all of which correspond to the same macroscopic state. Each of these points would represents some systems and these systems will have the feature that their microscopic structures are different but their bulk behaviour is same. **A random collection of such system, each of which corresponds to the same macrostate but has different microstructure, is called an ensemble.**

The great importance of the ensemble concept arises from the remark of *Gibbs* that the average over a period of time for a given system is equivalent to the average over the ensemble at one instant of time. An ensemble may be visualized as a collection of systems which are virtual copies of the real system i.e. macroscopically identical to real system but in different microstate. A microstate is defined as the specification of the individual position of phase points for each system or molecules of the ensemble. Individual system or molecule in the phase space is called phase point. A phase may be divide into cells and the phase point for any system or molecules may reside in one of the cells.

MAXWELL-BOLTZMANN'S DISTRIBUTION

The *Maxwell-Boltzmann* distribution establishes the distribution of macrocopic particles in phase space subjected to the action of arbitrary potential field. The *Maxwell-Boltzmann* distribution is applicable in classical system having interacting particles in equilibrium. We use this distribution if the number of phase points in a cell is very much smaller than the number of compartments.

Let us consider particles that are distinguishable. The number of microstates corresponding to any given macrostate is called thermodynamic probability of the macrostate and the probability for N particles placing into cells and the ith cell having n_i particles is,

$$W = \frac{N!}{\Pi\, n_i!} \qquad B(5)$$

For maximization of the probability, let us take log both sides

$$\log W = \log N! - \sum_i \log (n_i)!$$

Using *Stirling*'s approximation we get

$$\log W = (N \log N - N) - \sum_i (n_i \log n_i - n_i)$$

$$= N \log N - \sum_i N_i \log n_i \text{ . Since } \sum n_i = N$$

Differentiating the relation and using $d \log (W) = 0$ we get,

$$\Sigma \log n_i\, \delta n_i = 0 \qquad B(6)$$

Let the total energy of all n_i particles whose phase points lie in the ith cell is $E_i n_i$ and the internal energy of the system is then

$$U = \Sigma E_i n_i$$

The total energy of the system remains constant then,

$$\delta U = \Sigma E_i \delta n_i = 0 \qquad B(7)$$

We now use *Lagrange*'s method of undetermined multipliers to solve such equations $B(6)$ and $B(7)$. Thus for convenience let $\log \alpha$ be subtracted with $B(6)$ and along with $B(7)$ by multiplying β in such a way that

$$\sum_i (\log n_i - \log \alpha + \beta E_i) = 0$$

i.e.

$$\log \left(\frac{n_i}{\alpha}\right) = -\beta E_i \text{ and } n_i = \alpha e^{-\beta E i} \qquad B(8)$$

which is called *Boltzmann* distribution function of particles. A rigorous mathematical calculation shows that $\beta = \frac{1}{KT}$ and α identifies as the fugacity and is given by $\alpha = \frac{Nh^2}{V} (2\pi mkT)^{-\frac{3}{2}}$. V is the volume of the phase space and h is *Planck*'s constant.

Now the distribution function in general is expressed as

$$n_i = N\, e^{-\frac{E_i}{kT}} \qquad B(9)$$

However thus distribution cannot explain when there are large numbers of particles. It also cannot explain for microscopic particles such as electrons, protons, photon, phonon etc. These particles requires quantum distribution function such as *Fermi-Dirac* and *Bose-Einstein* distributions.

Some Typical Questions

C Appendix

1. Theory of Metal

(i) Describe the wave-particle duality with suitable examples.

(ii) Starting from ordinary wave equation in classical mechanics given by $y = A \sin\left(2\pi ft - \frac{2\pi}{\lambda} x\right)$ where all the notations have their own meaning, derive Schrödinger's equation in 1-dimensional form $\frac{\partial^2 \psi}{\partial x^2} + \frac{8\pi^2 m}{h^2}(E - V)\psi = 0.$

(iii) What is the significance of work function in metals?

(iv) How to you understand wave function? What are its significances in quantum mechanical studies?

(v) Starting from general solution of Schrödinger's equation given by $\psi(x) = A \sin \beta x + B \cos \beta x$ where $\beta = \left[\frac{8\pi^2 m}{h^2}(E - V)\right]^{1/2}$ prove that the energy of a particle confined in an infinite potential well is quantized.

(vi) What are eigenvalues and eigenfunctions?

(vii) Define and describe Heisenberg's uncertainty principle. How do you interpret this principle? Explain your views with suitable examples.

(viii) Using Heisenberg's uncertainty principle, prove that the minimum kinetic energy of a particle confined in an infinite potential well of width L is given by $E = \frac{\pi^2 \hbar^2}{2mL^2}$.

(ix) What are the significances of operators in quantum mechanical studies? Find the corresponding operator notations for energy momentum.

(x) Explain the behavior of particle confined to a potential box with finite barrier height. How is the particle wave function represented in the barrier region?

(xi) What is tunneling? Can you imagine the phenomenon identical to tunneling taking place in classical mechanics? How the wave function will be represented in either side of the narrow potential barrier that has been successfully tunneled through by the particle?

(xii) Show that for a linear solid, the energy is still quantized.

(xiii) Define number of states and density of states functions in quantum mechanics. Derive appropriate expressions for number of states and density of states.

(xiv) What is the application of Fermi function in quantum mechanics? Occupation index is highly temperature dependent. Justify.
(xv) Prove that Fermi energy in a metal is independent of temperature and depends only in its electron concentration.
(xvi) Using image charge method, derive the Richardson's expression for thermionic emission for Schottky effect.
(xvii) How the contact potential between two different metals with different Fermi levels can be employed to measure temperature of one of the metals?
(xviii) What is thermoelectric emf? Explain the Seebeck effect.

2. Free Electron Theory of Conduction in Metals

(i) Define lattice array, lattice parameter, packing density, Bravais lattices applicable to crystalline structures.
(ii) Find the packing densities of Si, Cu and Fe crystals.
(iii) Explain the procedures for identifying Miller indices for crystal direction and plane.
(iv) Explain the bonding in hydrogen molecule. What are bonding and anti-bonding molecular orbitals? Give reasons behind completely or partially filled bonding orbitals and vacant antibonding orbitals.
(v) Produce a neat sketch of Kronig-Penny model. Does this model correctly represent the electron within the crystalline structure?
(vi) What are the brilliouin zones?
(vii) Define and explain the effective mass of electron within a crystal. How do you understand negative and infinite effective mass of electron?
(viii) Starting from $PV = \frac{Nmv^2}{3}$ for gas in a container, derive the expression for mean thermal velocity of electron $v_{th} = (3kT/m)^{1/2}$.
(ix) Define drift mobility of electron, conductivity and resistivity applicable to metals.
(x) What are the factors that affect electrical conductivity in metals?
(xi) Derive $J = \sigma E$ for metals.
(xii) Electrical conductivity of metal is high although drift mobility of electrons in metal is low. Why? Explain.
(xiii) What are the factors that define the electron drift velocity in the metals? How temperature affects the electron drift mobility in metals? Is the same relationship applicable to semiconductors?

3. Conduction in Liquids and Gases

(i) What are the principal conduction mechanisms in liquids? Explain.
(ii) Explain the breakdown phenomenon in gases.
(iii) What are the principal conduction mechanisms in gases? Explain.
(iv) Explain the Townsend current growth mechanism.
(v) At normal temperature and pressure, gas is more likely to breakdown by an applied electric field than liquid dielectric.
(vi) How conduction starts in atmospheric air?
(vii) How frequently are the liquid and gaseous conductors used for current conduction purposes? Give some examples.

(viii) How the conduction in gaseous medium is different from that of solid metallic conductors? Explain.

4. Dielectric Materials

(i) How the macroscopic parameters are related to microscopic parameters in dielectric materials? Show.
(ii) Define dipole moment, polarization vector, polarizability.
(iii) How the electronic polarization takes place in an atom? Show that the magnitude of surface polarization charge density is equal to polarization vector.
(iv) How dielectric susceptibility is related to electronic polarizability?
(v) Derive the Clausius-Masotti equation relating polarizability with the permittivity.
(vi) Explain ionic polarization mechanism with necessary mathematical tools.
(vii) Give reasons for non-occurrence of orientational polarization in polar molecules under high frequency electric fields.
(viii) What are the parameters on which orientational polarizability is dependent? Explain with necessary derivation.
(ix) Explain interfacial polarization mechanism.
(x) Establish that polarization is frequency dependent.
(xi) Upon which parameters, does the power loss in capacitor depend?
(xii) Briefly explain the breakdown mechanisms in solid and gaseous dielectrics.
(xiii) What is ferroelectricity and piezoelectricity? What are similarities and differences between ferroelectricity and piezoelectricity?
(xiv) Name the field of application of different types of dielectric materials.
(xv) In your opinion, how does the polarization mechanism affect in the operation of an electrical system having many components in which polarization takes place during their normal operation? Are the effects same in both the ac and dc system? Explain.

5. Magnetic Materials

(i) Define magnetic dipole moment, atomic magnetic moment and magnetization vector.
(ii) Based on magnetization, differentiate between ferromagnetism, ferrimagnetism, antiferrimagnetism and paramagnets.
(iii) Explain the domain structure of ferromagnetic materials.
(iv) What is magnetostriction?
(v) Explain the magnetization and demagnetization of ferromagnetic materials.
(vi) What is the significance of hysteresis loop? Explain.
(vii) Mention the properties of soft magnetic materials.
(viii) Why ferrites are used for high frequency applications? Explain.
(ix) What is the reason behind use of hard magnetic materials for manufacture of permanent magnets and memories?
(x) Mention common properties of hard magnetic materials.
(xi) What are the common hard magnetic materials? How they are prepared?
(xii) What is the reason behind jerks in magnetization vector during magnetization process?
(xiii) Why pure Fe is not used in manufacture of ac machinery?
(xiv) Which is the most commonly used magnetic material for ac machinery? And, why?
(xv) Differentiate between hard and soft magnetic materials. Suggest which type of magnetic materials to use for the manufacture of power transformer cores and why?

6. Semiconductor Materials

(i) Justify the use of group IV materials as semiconductors. Are all the group IV materials semiconductors? Justify your answer.

(ii) Give properties of intrinsic semiconductors.

(iii) Describe the process of creation of electron-hole pair in intrinsic semiconductors.

(iv) What is the rationale behind assumption that the impurity concentration (either donor or acceptor) in extrinsic semiconductor is equal to the majority carrier concentration in the semiconductor material? Can this assumption be applied in all cases or are there any conditions attached to it? Explain.

(v) Define majority carrier, minority carrier, ionization energy in extrinsic semiconductors.

(vi) What is compensation doping? What are the differences in a n-type of semiconductor with a net carrier concentration (both majority and minority) that is equal to the same in compensation-doped semiconductor? Explain.

(vii) Which of the physical properties in the semiconductor compensation doping will considerably alter? Explain.

(viii) What is electron affinity in semiconductor? Give the reason for having the work function being greater than the electron affinity.

(ix) What are the mechanisms for generation of only electrons or holes in semiconductors? Explain.

(x) Despite having high mobilities compared to that of metals, semiconductors have low electrical conductivities. Give reasons for this.

(xi) Give physical significances of N_c and N_v.

(xii) Starting from $n \approx N_c \exp\left[-\dfrac{E_c - E_F}{kT}\right]$ and $p \approx N_v \exp\left[-\dfrac{E_F - E_v}{kT}\right]$, derive the expression for intrinsic concentration in semiconductor in terms of energy gap E_g, temperature, N_c and N_v.

(xiii) Where does the Fermi energy lies in semiconductors? How it is related to band gap in non-degenerate semiconductors? Explain with necessary derivation.

(xiv) What are degenerate and non-degenerate semiconductors? Does the Fermi energy in semiconductors always lay in the band gap? Give explanation to your answer.

(xv) What is minority carrier suppression in extrinsic semiconductors? Explain.

(xvi) What are the factors that affect most the conductivity of intrinsic semiconductors?

(xvii) Justify that the conductivity associated with minority carrier concentration can be neglected in extrinsic semiconductors.

(xviii) Describe the effect of temperature on resistivity of intrinsic and extrinsic semiconductors. Are the effects identical in both the cases? Give justification to your answer.

(xix) Differentiate between diffusion and drift mechanisms of charge carriers in semiconductor.

(xx) What is the impact of temperature on mobilities of charge carriers in semiconductors?

(xxi) How band bending occurs in semiconductors?

(xxii) Derive Einstein relationship.

(xxiii) What are the trap centers in semiconductors? Explain the indirect thermal generation and recombination mechanism.

(xxiv) How the EHPs generated by photogeneration behave after the source of illumination is switched off? Explain with necessary mathematical derivations.

(xxv) Define minority carrier diffusion length, minority carrier lifetime. How they are related to each other?

(xxvi) What happens when a n-type semiconductor is brought into contact with a p-type? Explain with necessary diagrams.

(xxvii) What is built-in electric filed and built-in potential? What will be the direction of internal field in the vicinity of the junction and why?

(xxviii) How is the built-in potential related to carrier concentrations in the semiconductors?

(xxix) Prove that the depletion width is mostly extended in to the region with lower impurity concentration compared to the region with higher impurity concentration.

(xxx) A junction is formed by joining n-type and p-type semiconductors with respective donor and acceptor concentrations of 10^{16} and 10^{18} per cm^3. How will the depletion region be extended in either region?

(xxxi) What are the factors that mainly affect the width of the depletion region in an abrupt pn-junction? Explain.

(xxxii) Explain with the energy band diagram the forward biased pn junction. Derive mathematical expressions for the same.

(xxxiii) What are the main factors that affect the current through forward biased pn- junction.

(xxxiv) Explain with energy band diagram how the charge carriers behave in a reverse biased pn junction.

(xxxv) What are the factors that mainly determine the magnitude of current through a pn junction? How does the temperature affect the current in reverse biased pn-junction? Can the current due to thermal generation overwhelm the reverse saturation current? Explain.

(xxxvi) What will be the affect of reverse biasing in pn junction on built in potential, internal field and depletion width?

(xxxvii) What is a Schottky junction? How do you understand the term Schottky barrier height? How is the current through the junction expressed during open circuit of the Schottky junction?

(xxxviii) What are the basic differences between Schottky junction and ohmic contacts?

(xxxviii) Expalin the process of formation of Ohmic contact with necessary energy band diagram.

(xxxix) What happens when a metal and an n-type semiconductor are brought into contact with the function of metal being greater than that of the n-type semiconductor? Explain with the energy band diagram.

(xl) Discuss whether doping always increases the conductivity or not.

(xli) A Si wafer has been doped with 10^{17} arsenic atoms per cm^3. Calculate the conductivity of the sample at 300*K*. Where will be the Fermi level with respect to the intrinsic Fermi level in Si? Calculate the conductivity of the sample at 400*K*. Use graphs wherever required.

(xlii) The sample given in question (xli) is further doped with 9×10^{16} boron atoms per cm^3. Calculate the Fermi level of the sample and conductivity at 300*K*. What type of semiconductor will result in?

(xliii) Produce a neat sketch of the dependence of the conductivity on the temperature as $\log(\sigma)$ versus 1/T and mark various critical temperatures and other relevant information.

(xliv) Sketch schematically the dependence of electron concentration in the conduction band on the temperature as log (n) versus 1/T, and mark the various important regions and critical temperatures. For each region draw an energy band diagram that clearly shows from where the electrons are excited into the conduction band.

(xlv) What is the average number of valence electrons per atom for a pair of Ga and As atoms and in GaAs crystal?

(xlvi) What will happen if Se or Te from group VI are substituted for an As atom in GaAs crystal? Explain.

(xlvii) What will happen if Zn or Cd from group II are substituted for an Ga atom in GaAs crystal? Explain.

(xlviii) Calculate the intrinsic conductivity and resistivity of GaAs at room temperature. The intrinsic concentration, electron mobility and hole mobility of GaAs are 1.8×10^6 per cm^{-3}, 8500 $cm^2V^{-1}s^{-1}$ and 400 $cm^2V^{-1}s^{-1}$ respectively at 300*K*

(xlix) For the semiconductor in question xlviii, where will be Fermi level with respect to the intrinsic Fermi level if the ionized donor concentration is 10^{15} cm^{-3}? Also calculate the conductivity of the sample. Take the electron and hole concentrations from above question.

(l) Consider an n-type silicon that is heavily doped and the electron concentration in the conduction band is 10^{20} per cm^3. Where will be the Fermi level? Can the expression $n_i^2 = np$ still valid to calculate the hole concentration? What type of semiconductor is this?

(li) A Schottky junction formed by Au and n-type semiconductor based on silicon with donor concentration of 10^{16} per cm^{-3} have a cross-sectional area of $1mm^2$. Calculate the theoretical barrier height from metal to semiconductor if the work function of Au is 5.1 eV and that the electron affinity of silicon is 4.01eV. And, what will be the reverse saturation current through the junction given the experimental barrier height of 0.8eV? Also calculate the current through the junction when a forward bias voltage of 0.3V is applied.

7. Semiconductor Materials Processing

(i) How is the pure silicon material obtained?

(ii) Why silicon needs to be purified?

(iii) What is the reason behind widespread application of silicon in semiconductor materials processing and manufacturing?

(iv) Explain the floating zone method of crystal growing.

(v) How a single silicon crystal is obtained by Czochralski growth method? Explain with a neat sketch.

(vi) Name the mechanisms of introducing impurities to the silicon semiconductor.

(vii) What is diffusion? State Fick's laws of diffusion.

(viii) Describe different diffusion systems. Compare between them.

(ix) Describe chemical reactions for obtaining boron trioxide which will eventually be the source of boron to be diffused in to the silicon semiconductor.

(x) What is the common source of phosphor that is doped in to silicon to form n-type semiconductor?

(xi) What are the common sources of phosphorous pentaoxide, which is necessary for donor diffusion in silicon? Give relevant chemical reactions.

(xii) How arsenic and antimony are obtained, which will ultimately be doped in to the silicon to form an n-type semiconductor?

(xiii) Explain the ion implantation process with a neat sketch.

(xiv) What are the advantages of ion implantation process over diffusion process?

(xv) Compare between diffusion and ion implantation process for impurity doping in to silicon semiconductor.

(xvi) Produce the Gaussian distribution resulting from ion implantation.

(xvii) Define lattice damage and annealing with reference to ion implantation process in semiconductors.

(xviii) How silicon and silicon dioxide are deposited by CVD method from silane gas?

(xix) What is epitaxial growth? What are the most commonly used compounds for epitaxial growth in silicon semiconductor?

(xx) Define photolithography, etchant, and monolithic fabrication.
(xxi) Describe the planar process of device fabrication with necessary sketches.
(xxii) Why metallization of contacts is necessary?

8. Superconductivity

1. What is the difference between perfect conductor and superconductor? Explain how do the electrical, thermal and magnetic properties of superconductors differ from those of normal conductors.
2. Explain the difference between type I and type II superconductors.
3. Explain the term critical magnetic field in a superconductor. How does the critical magnetic field vary with temperature in type I and type II superconductors?
4. What is *Meissner* effect? Explain how *Meissner*'s effect is complete for type I and incomplete for type II superconductors?
5. Derive *London* equation and explain the *London* penetration depth.
6. What is coherence length? Derive the relation of coherence length in terms of gap energy and Fermi velocity.
7. What is critical current? Prove that the critical current decreases linearly with the increase in applied field for a wire.
8. What does the equation B = Constant tell ? Is this equation adequate to explain superconductivity?
9. Write a short essay on the application of superconductivity.
10. Give qualitative description of BCS theory. Explain how does it account for superconducting state?
11. Expalin BCS ground state. How is it different from non - interacting Fermions?

Some Common Physical Constants

D

Appendix

Atomic mass unit	amu	$!.66054\times10^{-27}$ Kg
Avogadro's number	N_A	6.022×10^{23}/mol
Bohr magneton	β	9.27×10^{-24} J/T
Mass of electron	m_e	9.11×10^{-31} Kg
Charge of electron	e	1.602×10^{-19} C
Mass of proton	m_p	1.67×10^{-27} Kg
Electron magnetic moment	μ_e	9.28×10^{-24} J/T
Proton magnetic moment	μ_p	1.41×10^{-26} J/T
Nuclear Magneton	μ_N	5.05×10^{-27} J/T
Boltzmann constant	k	1.38×10^{-23} J/K
Absolute permeability of vacuum	μ_O	$4\pi\times10^{-7}$ H/m
Absolute permittivity of vacuum	ε_O	8.85×10^{-12} F/m
Planck's constant	h	6.624×10^{-34} Js
Reduced Planck's constant	$\hbar$	1.055×10^{-34} Js
Speed of light	c	2.9979×10^{8} m/s
Atomic density of Si	ρ	5.0×10^{22}atoms/cm^3
Intrinsic carrier concentration of Si	n_i	1.5×10^{10}cm^{-3}
1 Angstrom	Å	10^{-10} m
1 eV		1.602×10^{-19} J
1exa	E	10^{18}
1peta	P	10^{15}
1tera	T	10^{12}
1giga	G	10^{9}
1mega	M	10^{6}
1atto	a	10^{-18}
1femto	f	10^{-15}
1pico	p	10^{-12}
1nano	n	10^{-9}
1micro	μ	10^{-6}

E

Appendix

Bibliography

1. Colclaser, R A, Diehl-Nagle, Sherra, Materials and devices for electrical engineers and physicists, McGraw Hill, New York, 1985.
2. Kasap, S O, Principles of electrical engineering materials and devices, McGraw Hill, New York, 2000.
3. Jaeger, R C, Introduction to microelectronic fabrication volume V, Addison-Wesley Publishing Company, Massachusetts, 1988.
4. Naidu, M S , Kamaraju, V, High voltage engineering, 2nd edition, Tata McGraw Hill, 1997.
5. Dekker, A J, Electrical engineering materials, Prentice Hall of India P Ltd, New Delhi, 2000.
6. Ashcroft and Mermin; Solid State Physics, Saunders College Publishing, USA, 1988.

Index

About the Authors

Dr. Pokharel is a Professor of Physics at Institute of Engineering, Pulchowk Campus. He is a Senior Associate of the Abdul Salam International Centre of Theoretical Physics, Trieste, Italy. His academic qualifications include M. Sc. (TU) in Solid State Physics, M. Sc. (UK) in Materials Science and Ph.D. (India) in Materials Science and Technology. He has about 20 years of experiences in research and teaching in undergraduate and graduate levels. He was visiting scientist in 1998 at the Centre for Advanced Technology, Defense Ministry, Indore, India. His research contributions include "Synthesis, Structure and Phase Transition Behaviour of Ferroelectric and Antiferroelectric ceramics as well as Polymers materials. He has contributed more than dozen of research papers in national and international reputed Journals. He has conducted several national and international materials science projects. Under his guidance about half a dozen of M. Sc. (Physics) students have conducted their thesis works.

Mr. Karki is currently working as Lecturer and Program Coordinator, M. Sc. Power Systems Engineering Program at the Department of Electrical Engineering, Pulchowk Campus. He completed M. Sc. in Electrical Power Systems (Hons) in 1995 from Belarussian State Polytechnical Academy, Minsk, Belarus. After the completion of the study in Belarus, he joined Institute of Engineering, Tribhuvan University as a lecturer and taught at Purwanchal Campus, Dharan from 1995 to early 1999. Then he moved to Pulchowk Campus where since then he is teaching at both the undergraduate and postgraduate level.

He has supervised many undergraduate projects and Master's theses at the Institute of Engineering. He was a Guest Research Fellow at Osaka Sangyo University, Japan in the year 2000 and has participated and presented papers at many international conferences. He is member of IEEE, Nepal Engineers Association and a registered professional engineer in Nepal. His research interests include Power System Planning, Operation and Control, Power Sector Restructuring, Distributed Generation, Mini Grid, High Voltage Engineering, Renewable Engineering.